Awesome Green

by

Carl Tant

Biotech Publishing
A Division of
Plant Something Different, Inc.
Angleton, TX, USA

Awesome Green

By CARL TANT

Published by Biotech Publishing
P.O. Box 1032
Angleton, TX 77516-1032

Printing 10 9 8 7 6 5 4 3 2 1

Library of Congress Catalog Number: 93-073250
ISBN 1-880319-05-5 US $17.95 Softcover

Publishers Cataloging Data:
Author. Title
Biotechnology, plant
Foods, transgenic
Genetic engineering
Medicine from plants
Industrial products from plants

Cover by Tammy K. Crask

Table Of Contents

Photo Credits

Dr. Charles Arntzen, Texas A&M University Institute Of Biosciences And Technology

Mick Farms

The Monsanto Company

Mycogen, Inc.

Dr. Craig Nessler, Dept. Of Biology, Texas A&M University

Colinda Roden

Other photographs by the author and illustrator.

Trademarks

Cell Cap - Mycogen, Inc.

Clorox - The Clorox Company

Flavr Savr - Calgene, Inc.

Roundup - The Monsanto Company

Synthephytes - Plant Something Different, Inc.

Transwitch - DNA Plant Technology Corp.

VineSweet - DNA Plant Technology Corp.

WordPerfect 5.1 and WordPerfect For Windows - WordPerfect Corp.

Zephiran™ - Winthrop Laboratories, Inc.

Acknowledgements

The input of many is a vital necessity for a book of this kind. The author and publisher gratefully acknowledge the assistance of the university faculty mentioned and the many corporate scientists and communications department personnel who provided information and the pictures listed in photo credits.

Others provided invaluable help and deserve recognition:
- Renee' Walker for library research assistance and some text entry in WordPerfect 5.1™.
- Colinda Roden for library research assistance and proofreading of early versions of the manuscript.
- Suzanne Muecke, English Chair at Angleton Senior High School, for wielding her red pen on various grammatical sins.

A very special thanks is due Amy Harris for exercising her talent and patience in page layout and preparation of final printing copy in WordPerfect for Windows™.

To ...

My sister, Nila Tant Middleton, who read and understood the
manuscript for this book. She knows everything about real
estate, but nothing about science, so if she could
understand it, everyone can.

Foreword

The sciences of plant biotechnology have developed so rapidly that the general public is largely unaware of the great promises they hold. Even science teachers at the preparatory levels and non-specialists in college have difficulty in keeping up with the latest developments.

Unfortunately, the biotechnology industry in general has often failed to keep the public informed and aware of the values of current research. Much of their public relations work seems to have been directed to financial institutions in efforts to obtain funding. The public has frequently developed its perception of plant biotechnology from science fiction and from those who oppose this type of research.

In the academic world, some of the problem may be the lack of experience of the research professor in communicating with laymen. The author received a cordial and helpful reception from those whom he contacted to obtain more information. They seemed relieved to learn of the author's technical background which permitted them to discuss their research in scientific terms. Many also seemed to marvel at the idea that it might be possible to translate this into an understandable book. Their difficulty with lay communication is not necessarily a failure on the part of the professors. Their life and future is at stake in the "publish or perish" philosophy of many major universities. The pressures of their work have simply precluded many from contacts below the graduate student level.

Hopefully, we have cleared up the mysteries.

No information has been taken from the popular press. All subjects contained herein have been presented in peer-reviewed scientific literature.

This book makes no pretense at being a thorough review of the current literature. It does attempt to present some of the major advances of plant biotechnology in terms that the average reader can understand. The facts and the promises of this new science for society as a whole are far more excitingly intriguing than the science fiction and scare stories.

In a recent editorial in **Genetic Engineering News** (June 15, 1993) publisher Mary Ann Liebert commented, "The underscoring point is that early public education is essential if new products are not to experience unreasonable delays due to activists who may not completely understand new technologies and ramifications. ...last, but not least, biotechnology's impact on health, agriculture, and environment can be a stimulating proposition for young minds contemplating the world's problems." The author and publisher hope this book will provide some fulfillment of the needs she addressed.

Finally, why did we choose this title? Quite simply the answer is that young people often use AWESOME for something which is so great that it almost defies description. Such is the new green world.

Part I

Fables, Fears, And Facts

A Prophecy Of Doom ...

Chapter 1

A Prophecy Of Doom ...

Texas

The building seemed small and almost insignificant, often shadowed by the giant towers of the hospitals and other professional buildings in the sprawling medical center. The sign above the door read "Pediatric Clinic."

Inside, the young couple sat close to each other in the waiting room. Their faces were etched with lines of worry and sleepless nights. Subconsciously clasping hands, they stared intently at the examination room into which they had earlier carried their ten-year-old daughter. Their look at the door silently willed a miracle to happen. The clock calendar on the wall showed March 6, 2009.

Within the examination room the pediatrician and the geneticist stood silently looking at the frail child who lay on the examining table. She was quietly dozing, seemingly at peace, giving lie to the appearance of her body. Her weak limbs were showing early signs of degeneration of muscle structure. Her back curved, obviously indicating extreme weakness of the supporting muscles. Small areas of the outer layer of skin had sloughed off at the slightest touch.

An electronic beep from the computer screen at the head of the examining table caught the doctors' attention. They looked intently, fearing, but deeply knowing that the laboratory results would confirm their diagnosis. The DNA analysis came up on the screen. *The doctors glanced at each other, their eyes expressing the frustration of hopelessness. The nurse held back the tears that welled up in her eyes while the doctors silently turned and started toward the waiting room.*

As the doctors approached, the parents read the dreaded verdict in their faces even before they started to speak.

The pediatrician began, "To come to the point, because there's no easy way to say this–nothing we say is going to make it much easier. Your daughter has what we refer to as a genetic development regression. The genes that control development of her body and even those which maintain what has already formed are in a state of failure."

The geneticist added quietly, "Something from outside has invaded her body and in effect, turned off a giant switch which maintained normal inherited control. At this time, we don't know what this substance or thing is."

The pediatrician read the last glimmer of hope and pleading in the parents' eyes. Silently, he shook his head. "I'm so deeply sorry, but there is just nothing we can do in the state of our present technology. I can only assure you that your daughter will not suffer during the few weeks of life that remain. This is a new disease–the first cases began to appear just a few years ago. They're starting to increase now. We're still searching for something all these children have in common, but so far we haven't found it."

"The only consolation I can offer is that you might find some outlet for your grief in providing information which could help other children in the future. That would at least help give your daughter's life a magnificent function. Whenever you feel up to it, there are a few questions we would like to ask about events around the time of conception and during early infancy. It won't take long. Please do let us know when you want to help."

"Now!" The mother spoke intently. "Every day could make a difference for another. What do we need to do?"

The geneticist spoke. "Researchers are starting to home in on an idea that the underlying cause might be an unusual DNA fragment or perhaps a virus-like particle or something similar that invaded the egg or sperm around the time of fertilization, and slowly brought about changes in the normal development control genes. There's one little thread which is starting to show up in many of these cases. Think back to the middle 1990's. Did either of you use and eat any of the genetically engineered fruits and vegetables that came on the market during that time?"

The mother and father thought for a moment and the answer came from both.

"Yes."

Arizona

The rancher sat down at the huge desk in his study and picked up the sweat-stained note left by his foreman. It had obviously been hurriedly scribbled, probably while in the saddle:

The rancher angrily banged his fist so hard on the desktop that everything on it jumped and moved. He reached over to straighten things and noticed the date on his calendar - April 14, 2009.

Leaning back in his chair, he continued to stare at the date. So much had changed in the last 40 years while he operated this great ranch inherited from previous generations. At the same time, so much of the life he had chosen remained unchanged, virtually as it was a hundred years before. He thought of how he had studied and embraced the new technologies that had brought cattle breeding to almost becoming a predictable science. More lean beef–less fat, lower cholesterol. The more healthy red meat knowledgeable consumers demanded today. He closed his eyes and began a mental list of what could possibly have suddenly gone so horribly wrong. No herbicides were used on the ranch. No new insecticide sprays–only the old time-honored dips for fleas, ticks, and other external parasites. No food additives. No visible change in range conditions.

He arose from his chair, walked to a bookcase, and extracted his old college genetics textbook. He opened the cover to the copyright page and observed ©1968. His finger flipped through the book to reveal a proud diagram of the double helix of DNA. He stared at it for a moment and turned a few pages over to the end of the chapter, there reading a glowing paragraph which predicted that in the not-too-distant future man would be able to apply the new technologies of genetics to bring about wondrous and beneficial changes in all forms of life. Disgustedly he snapped the book shut and shoved it back on the shelf. The scientists then were like children with a new toy, enamored by its brightness and glitter, but not yet having learned how to play with it.

He returned to his old comfortably worn leather chair. Tilting back, he thought again. There was simply no explanation. They had done everything–checked for water pollution, checked background radiation, etc. Nothing was beyond normal limits.

Then the thought struck. Yes, there was one thing that had changed. Three years ago they had planted seed of a new hybrid grass in different areas of the vast ranch, but surely that couldn't be it. The grass really wasn't so different; in fact, it was actually the native pasture grass upon which cattle had grazed for over 150 years. The only thing different about it was that the genetic engineers had taken it a few years earlier and succeeded in combining into it a gene which produced a higher protein content and another which slightly increased its drought tolerance beyond the natural level. Otherwise, it seemed the same; that couldn't possibly be an explanation, could it?

South Carolina

Across the nation from the rancher, a Parks and Wildlife ranger swore to himself as he attempted to extract a rabbit which had become entangled in the stickiness of a gigantic Cape Sundew plant. Finally freed, the rabbit went on its way, pausing only to rub on grass and lick, attempting to rid itself of the gummy residue still adhered to its fur.

The ranger looked at the sundew plants around him. He thought, "These are even bigger than last year. Some of them are almost two feet tall. What's happened?" he wondered. "Just a few years ago this little tiny plant required highly specific conditions for growth and got only an inch or two tall." The sticky material it exuded served to capture unwary insects and then it excreted some enzymes which partially digested them so that they became part of the plant's nitrogen supply. Harmless, very limited in its range, its only value was that of a carefully cultivated ornamental curiosity which some people took great pride in being able to grow out of its natural environment.

Now they seemed to be spreading everywhere. As he turned back toward his pickup, the ranger was struck with the bemused thought, "If they keep on at this rate, in another year or two, I'll

*have to be more careful where I walk. Wouldn't it be something
to get stuck on one of those things!"*

*Thinking back to his college biology courses, he muttered, "Well,
I guess this is what they were talking about when they lectured
on mutations. I wonder how."*

Iowa

*The basic processes of grain production had changed little
between the last years of the twentieth century and the first
decade of the twenty-first. The farmer parked his dusty pickup
in front of the county agriculture agent's office. Getting out, he
reached in the back and carefully avoiding its sharp spines,
picked up several feet of a vine. Mixed in with the green foliage
were a number of large bright flowers. He carried the vine with
him into the county agent's office, handed it to him and asked,*

*"What is this thing? It's all over the place. If it keeps on
growing like it is now, it's going to tangle up my corn harvesting
equipment."*

*The county agent examined the plant, consulted a couple of
reference books, and turned to the farmer. "Where did you say
you got this?"*

*The reply came, "In my cornfields. I saw one or two of these
last year; now they're all over the place, growing like wild.
What is it?"*

"Well, it looks like a vine known as Bougainvillea *in the south,
but it doesn't grow here. It can't survive our winters. Some
people cultivate it in the upper part of the south. In areas closer
to the Gulf of Mexico, it's grown as an ornamental where
freezing temperatures don't occur. But again, it doesn't grow
here."*

The farmer glared at the agent and then again stared at the plant. "It is growing here, and it's going to wipe out my crop."

The county agent followed the farmer to his cornfields. At a momentary glance, his surprise turned to shock and disbelief, and then to a realization of the horror of reality as he observed the prolific vine bending over corn plants with its weight and almost completely smothering other stalks that were approaching maturity and harvest.

"It has to be a Bougainvillea. But I can't believe what I see. Even if it were to grow through the summer, our season is too short for this kind of growth, and it would be killed by the winter. Frankly, I don't know. I'll call the University and get them to send some experts to see this. They won't believe my identification. This can't be."

But, it was.

... A Reality Of Eden

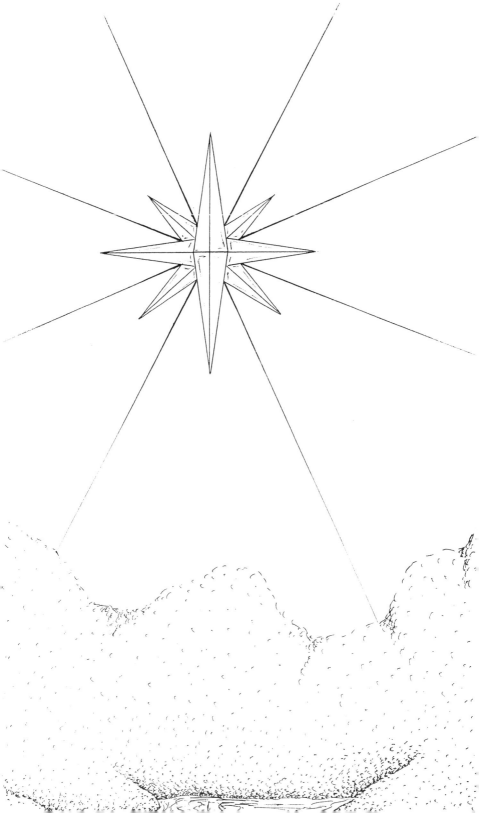

"The Reality"

Fortunately, the stories you just read are fiction. They do not even deserve the dignity of being called science fiction--fiction, yes; science, no.

Unfortunately, they are typical of ideas suggested by uninformed anti-technology do-gooders who oppose advances in plant biotechnology for various reasons. Also, unfortunately, they represent the very real fears held by many individuals who do not understand the realities of plant science today. The rest of this book is devoted to making the realities meaningful to those who lack a technical background in this field. No matter how educated and competent a person may be in his own specialty, the rapidly developing and transforming areas of plant genetics and technology probably represent a vast unknown. The real world of the new plant biotechnology is far more intriguing and exciting than the fiction and the scare stories--because it is real. It is a story of almost unbelievable potential for the betterment of mankind. It is truly the beginning of a world of awesome green.

We shall start our explorations by examining a few of the major specialized techniques that will bring this new world to reality. With an understanding of how these things are done, we shall then explore some of the most fascinating recent developments.

These are many - far too many to include them all in this volume. Selection has been difficult, but we have tried to pick those which have the widest appeal or are near commercial availability. Others were selected because they represent landmark technology or are unique in some special way.

Chapter 2

From The Old To The New

The green world is one which has changed, but at the same time has been unchanging. Many of today's plants differ little from their ancient ancestors. The ferns, mosses, cicads, horsetails, and other varieties of today are very similar, in some cases identical, to their ancient forms. The dominant forms have evolved in response to other changes in the world, particularly those of climate. Gone are the ancient forests of ferns and their allies which once covered much of the earth before it underwent change from tropical to temperate. In many instances, the remaining forms represent only a few survivors of once magnificently pre-eminent species. The beautiful maidenhair tree (*Ginkgo biloba*) is today the only survivor of its once vast genus. Variation has been inherent, and new species have developed as the reproductive processes of plants changed from simple asexual ones to spore formation to the complex sexual matings of modern flowering plants.

Man's dependence on the green kingdom has always brought about a fascination exceeding the basic need. Somehow the primitive gatherers learned to distinguish the edible from the harmful.

Later, other uses were discovered. Plants became a source of shelter, clothing, and eventually provided the accouterments of civilization such as paper.

Somewhere along the line early man also learned that certain plants had powers to heal his ills and pains. That slowly developing knowledge sometimes took strange turns such as the

promulgation of the Doctrine of Signatures which held that a
supreme being had placed plants on earth for the benefit of man
and had used their shapes to give clues to their uses. If a plant
or plant part resembled a part of the body, then it was considered
to be useful for treating that part. Hence, names such as
liverwort developed. Because this small insignificant bryophyte
vaguely resembled the shape of the liver, (Figure 2.1) it was

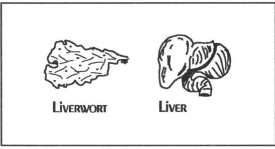

Figure 2.1

thought to be useful in treating
ailments of the liver. These were
many, for at that time if physicians
were unable to make another
diagnosis, the problem was usually
blamed on the liver because it was
the largest organ in the body. This
idea reaches its zenith today in the
case of ginseng root which is
believed by many to cure various
ills, stimulate sexual function, and
prolong life. A twisted root like
the one in Figure 2.2 which has the
shape of a whole man brings by far
the highest price.

Figure 2.2

From this somewhat tenuous and shaky start, plant science began to develop. Its history is dotted with advances brought about by a few whose names are known to almost every student: Carolus Linneaus, the Swedish botanist who first satisfactorily categorized plants and laid the foundation for our modern biological classification system; Gregor Mendel, who discovered the basic principles of genetics hidden in the peas of his monastery gardens; George Washington Carver and Luther Burbank who laid the foundations for modern breeding and improvement techniques.

Thousands of other less well-known individuals contributed vital links in the chain supporting modern plant biology. While some of their ideas seem primitively simple in the light of present sophistication, they nevertheless provided the necessary basics. For example, the Greek philosopher Theophastus some 300 years BC wrote in his **A History Of Plants**: "In some plants root and leaves germinate from the same point and in some from opposite ends of the seeds. In the former the seeds plainly have two lobes and are double while in the seeds being in one piece the root grows a little before the shoot. Barley and wheat come up with one leaf, but peas and beans with more leaves." Today, we refer to Theophastus' division as the two main groups of seed plants-- the monocots and dicots.

Another example of a major contribution by a virtually unknown name was the first accurate description of sex in plants by Rudolf Jakob Camerarius, who in 1694 wrote in his **De Sexu Plantarum Epistola**: "In plants no production of seeds takes place unless the anthers have prepared the young plant in the seed. It is thus justifiable to regard the anthers as male and the ovary with its style represents the female part. ...When I remove the male I never obtain the perfect seeds but only empty vessels which finally fall desiccated...."

And so it went. Bit by bit, piece by piece, the puzzle was assembled and the foundations put in place over thousands of years. By the late 20th century plant science was ready for its next gigantic step into the new world we call plant biotechnology. There has not yet been time for the names of the major contributors to become well known, but history will undoubtedly record their advances with honor.

Even this record is not complete. As we come to the end of the 1900's, the advances are a culmination of sometimes seemingly small increases in knowledge produced by countless hundreds of scientists and technicians working with intensity and honor in government, education, and industrial laboratories throughout the world. Together they have provided the tools and techniques which we shall explore in the next chapters.

Chapter 3

Time Out - Back To The Basics

Don't you just hate books that ramble on and on and on and on and on about the same old basic stuff you already know when you bought the book to learn something new? If you remember all you were supposed to have learned about plants in 7th grade Life Science, then skip this chapter and go on to the next. On the other hand, if you are saying to yourself "I don't remember because ..."

(Pick the best choice:)

    A. I was busy reading the comic book hidden in my text.
    B. I was busy passing notes to my friend in the next aisle.
    C. I was busy thinking about how to get that good looking girl/guy in my next class to notice me.

Read it and renew your memory.

The Plant Structure

You probably do remember the four major plant organs:

1. **Roots:** which anchor the plant in the soil and help it obtain water and dissolved nutrient through the root hairs.

2. **Stems:** which hold and distribute the leaves to obtain light.

3. **Leaves:** which carry out the process of photosynthesis by converting light energy into chemical energy.

4. **Flowers:** which are the reproductive parts of higher plants.

The functions listed above are the primary ones, but sometimes there are others. For example, roots such as a carrot, may also serve for food storage. Even stems, like the tuber of an Irish potato, may be modified for that same purpose. Other stems, such as the runner of a strawberry, also serve a reproductive function.

Roots, stems, and leaves which have no direct function in the reproduction of higher flowering plants are often referred to as the vegetative parts; their components are sometimes called the somatic tissues. The flower is the reproductive organ of higher plants. A complete flower consists of the sepals, petals, stamen, and pistil as shown in Figure 3.1. The stamen is the male

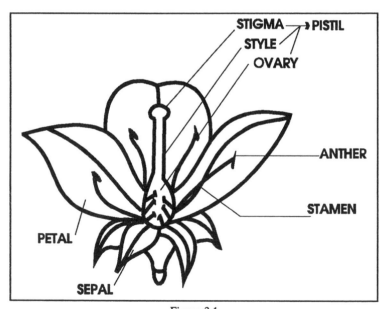

Figure 3.1

reproductive structure consisting of a stalk which supports the anthers on which are borne the pollen grains. The pistil is made up of the stigma, style, and ovary. Within the ovary is found the ovule or egg. A flower which does not have all of these parts is referred to as incomplete. If both male and female reproductive structures occur in the same flower, it is described as being perfect; if only the organs of one sex are present, the flower is called imperfect. While most plants have perfect flowers, there are many, such as squash, which produce separate male and female flowers. There are even some species in which separate male and female plants occur.

Reproductive Processes

All the lower non-flowering plants reproduce asexually by multiplication of vegetative structures, or produce special asexual reproductive cells such as spores. Even in these, there may be a sexual reproduction stage usually called the gametophyte generation. Even many of the flowering plants are capable of some type of asexual reproduction in addition to their sexual processes. A common example of the latter is an African Violet which will often produce a new plant from a leaf structure which contacts the soil. Another example is the popular house plant *Kalanchoe glossofeldiana* (Mother of Thousands) which is capable of producing hundreds of new plantlets along the margins of its leaves. When these have developed rudiments of all the plant's organs, they fall off, root in the soil, and grow into new plants. One of the most common means of propagating many plants is by taking cuttings from the stem and placing them in the soil where they will start to form roots and quickly develop into a new plant.

Sexual Reproduction

The process of sexual reproduction in plants is not appreciably different from that in animals. The pollen grain, when in contact with the stigma, produces a sperm which fertilizes the ovule. Following fertilization, the fertilized egg, now called a zygote, begins to divide and produces a mass of cells. As in the animal embryo, different groups of these cells begin to take on special characteristics and develop into the primary plant organs. In most plants, this embryonic structure becomes associated with various food reserves and protective materials which, along with the embryo, make up the seed. Once everything is in place, there is a loss of water from all the structures contained in the seed, and the embryo goes into a state of dormancy until conditions become right for growth.

Plants, like all other living things, have their heredity information in the sequence of the building blocks of the large chemical molecule known as deoxyribonucleic acid - DNA. The DNA is contained in structures known as chromosomes. The chromosomes exist as pairs of similar ones in all somatic cells. The number of chromosomes differs in various species of life, but a specific number is characteristic of each species. Every cell of the body has the complete genetic code; however only those parts applicable to that specific cell are turned on. The others are not expressed. In other words, the genes which have the code for the shape of your ears are kept turned off in your skin cells. As an early step in the reproduction of a cell, the number of chromosomes is doubled. Then the cell divides and each resulting cell has the same genetic code as the original. The obvious question that arises at this point is "So what happens in sexual reproduction where the fertilized egg has chromosomes from both parents? Does the number of chromosomes double in each generation?" The answer to the question is no and is explained by the mechanism shown in Figure 3.2.

In the process of developing the sperm and egg, the chromosome number is halved. Only one member of each pair of chromosomes goes into each sperm and egg. This is sometimes described as the haploid condition or *n* number of chromosomes. At fertilization, the similar chromosomes again form pairs to give the diploid condition which is characteristic of all somatic tissues. Thus, the total chromosome number stays constant from one generation to the next.

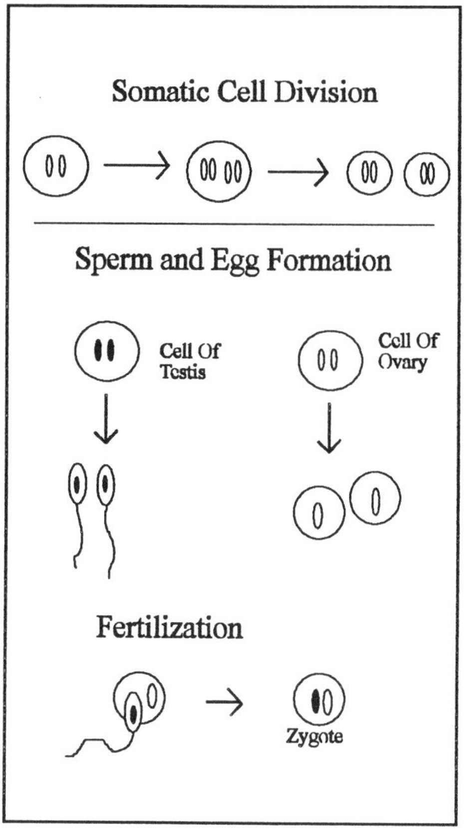

Figure 3.2

The ultimate end of this is that every cell which makes up the plant contains all of the chromosomes and consequently the entire genetic information in the form of the DNA which is the code for all the plant's characteristics. This condition is sometimes described as totipotency. Although the complete genome for every trait is present in every cell, only the codes concerned for that particular kind of tissue are switched on. The others are normally rendered non-functional, but can be artificially switched on through the use of hormones and other chemical or physical means. Some of these are discussed in the following chapters.

There is one other pair of terms which one must know in order to understand any discussion of biotechnology. These are *in-vitro* which literally means "in glass." It refers to processes which occur under artificial conditions in the laboratory in contrast to *in vivo* which means "in life." In other words, it refers to processes which occur in an actual living animal, plant, or microbe.

Part II

The Tools And Means

Chapter 4

Cloning

Just Another Form Of Propagation

The laboratory propagation of plants has been known by many names. Words used have included cloning, and an early term of orchid fanciers, mericloning. Particularly, the term "tissue culture" has been applied and misapplied to many types of laboratory work with plants since the initial attempts of Haberlandt in 1902, to grow plant cells independently of the parent plant. Technically, "tissue culture" should refer only to the growth of tissues made up of a limited number of cell types. The growth of organ systems such as the shoot or root should properly be referred to as "organ culture." The production of whole plants is most meaningfully described by the word micropropagation. Sometimes this is expanded to *in vitro* propagation. "Cell culture" should be used only to describe the multiplication of individual cells without organization into complex tissues.

The first real promise of success with *in vitro* multiplication came in 1934, when R. P. White for the first time succeeded in growing plant cells in artificial media. Many researchers contributed to the knowledge of the process for about thirty years. Finally, commercial applications began to come in the 1960's and 1970's, when Murshige and others developed nutrient media which provided for the rapid multiplication of many different species. By the 1980's these processes had been adapted to the commercial production of millions of plants. Not all species are adapted to this type of technology; for example, the woody plants as a rule have been much more difficult to propagate *in vitro* than have the herbaceous plants.

The ultimate purpose of micropropagation is to achieve the multiplication of plants on a large scale under laboratory conditions. The process is primarily useful for the production of varieties that are difficult, slow, or expensive to propagate by ordinary practices. Many plants, particularly desirable ornamentals, do not reproduce from seeds.

Another application which is assuming great importance is the multiplication of new or modified varieties from a very limited amount of starting material. Often, a single plant is all that is available.

Micropropagation is often the only way to achieve production of large numbers of such plants. However, all of the laboratory processes are relatively expensive and require special facilities along with highly trained personnel. Thus, the processes are not economical for species which can be easily reproduced by cuttings or seed. The general procedure for microprogation is shown in Figure 4.1.

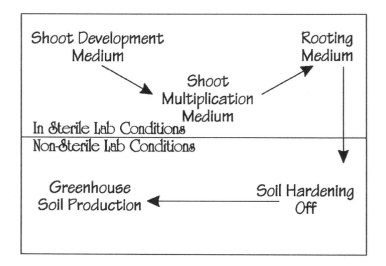

Figure 4.1

Physical Facilities

Most of the nutrient media used for *in vitro* propagation have a high sugar content. That makes an open invitation for mold growth. As a result, one of the major problems in micropropagation is the prevention and control of mold contamination. This is accomplished through the use of strict aseptic technique in rooms ventilated by filtered air. Mold spore removal is usually obtained by the use of HEPA (high efficiency particulate air) filters. In many installations the technicians doing the work utilize surgical gowns, masks, and other sterile attire. The filtered air within the facility is maintained at a positive pressure relative to the outside. Inoculation and various manipulations of culture media and other materials are usually done in special hoods or on bench tops facing HEPA filters. The incubation areas are equipped with fluorescent or mixtures of fluorescent and incandescent light designed to produce intensities ideal for the species being grown.

Heat build-up in such a facility can be a major problem. Higher than normal air conditioning requirements can usually be expected. In large incubation areas, the ballasts for fluorescent lighting may have to be located outside the room. The expense of this can sometimes be partially offset during winter months by utilizing the heat produced by the ballasts to supplement the regular heating equipment.

Figure 4.2 shows one possible arrangement of a micropropagation laboratory.

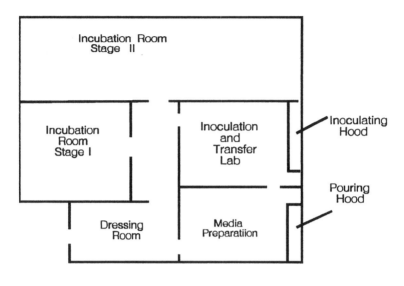

Figure 4.2

Plant Multiplication

Theoretically, any plant cell, tissue, or organ can be used as the starting material for plant development. From the practical standpoint, the most likely source is usually apical meristem. It is composed of rapidly multiplying cells at the growing tip of a shoot. Other structures are sometimes suitable. Plant material utilized to start a culture is usually referred as an "explant" regardless of the organ from which it was obtained. Ideally, the explant can be other micropropagated material already being grown aseptically. If sterile plant material is not available, the next best source is a plant grown aseptically from sterilized seeds. It is much easier to sterilize the outer surface of a seed then to sterilize mature plants.

The chances of obtaining only callus growth are much greater if nonmeristematic materials are utilized. Plant callus is similar to a callus that develops on your hand or foot. It is a special protective tissue formed in an area that is being injured by a cut or abrasion. The callus, or whatever tissue is produced, can be manipulated by controlling the hormone content of the nutrient media to bring about the development of shoots. Once these have been obtained, they can be placed into a nutrient media containing hormones known as cytokinens that will initiate the production of side shoots. After sufficient shoot growth has occurred, a rooting hormone will initiate root development. Finally, the plant is removed from the sterile nutrient and placed into a growth medium. Such plantlets are removed to a special area of a greenhouse to give them a chance to "harden off." Extreme care must be used during the first few days to avoid excess light and to maintain a high humidity because most plantlets developed *in vitro* do not have protective cuticle present on the leaves. Development of normal cuticle requires a week to ten days. After this period, the plant is like any other and can move through the standard processes of growth to the desired sale size.

Figure 4.3 illustrates the steps in micropropagation from a meristem explant.

Nutrients

Many different nutrient media have been devised for shoot multiplication. The most commonly used ones can now be obtained in a convenient dehydrated form from many laboratory suppliers. All of the nutrients used for *in vitro* propagation include macro- and micro- nutrient mineral salts, sucrose, vitamins, and hormones. A typical example is the Murashige and Skoög shoot multiplication medium A. The formulation for its minerals and vitamins is shown in Table 4.1. Sugar and hormones would be added because the shoots are too young to make their own. There is really nothing exotic about it.

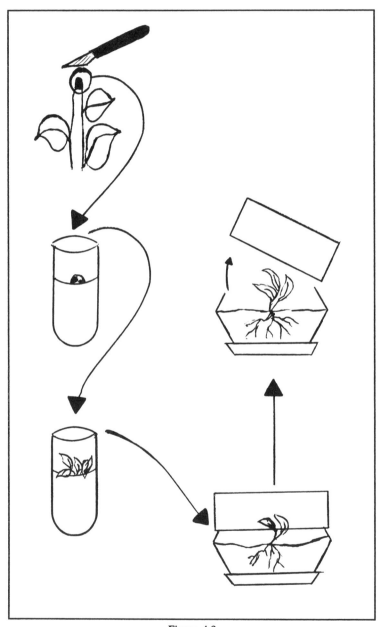

Figure 4.3

Compare the mineral content with that of your favorite houseplant or garden fertilizer!

Non-Meristem Sources

As noted above, if sterile micropropagated starting materials are not available, other plant sources can be utilized.

Seed is the next best starting material. It should be fresh and properly packaged. If dirt or chemical residues are visible, the seed should be washed thoroughly in a non-toxic mild detergent before starting sterilization. A typical protocol for sterilizing seed is shown in Figure 4.4.

If mature plant material must be used to start the process, it is usually sterilized by a similar procedure. If possible, exposure of the plant to full sunlight for a day or two before starting will reduce surface mold count. Washing with a mild detergent solution will also help. The chances of obtaining sterile cultures are better if only meristem tissue is used after the outer coverings are removed.

Another approach is the use of cell cultures as starting material. Frequently, shoot multiplication may not occur from the types of explant described above, but callus tissue will develop. The sterile callus cells can sometimes be removed from the original culture and placed in a nutrient containing hormones which will permit or promote embryogenesis or direct organogenesis. In many cases, techniques of hormone manipulation are utilized to deliberately obtain a large mass of callus tissue in order to speed up the whole process.

Table 4.1

Muraghige And Skoög Shoot Multiplication Medium

Chemical	Quantity (mg/L)
Macronutrients	
Ammonium nitrate	1650
Potassium nitrate	1900
Calcium chloride dihydrate	440
Magnesium sulfate heptahydrate	370
Ammonium sulfate	—
Sodium biphosphate monohydrate	—
Potassium phosphate, Dibasic	170
Micronutrients	
Manganese sulfate monohydrate	40
Copper Sulfate pentahydrate	0.08
Sodium Molybdate(VI)	1
Boric acid	12
Zinc sulfate heptahydrate	8
Iron EDTA	0.4
Vitamins	
Thiamine Hydrochloric Acid	400
Riboflavin	40
Nicotinic acid	40
Myo-Inositol	4000
Pantothenic acid	40

Figure 4.4

Seed Sterilization

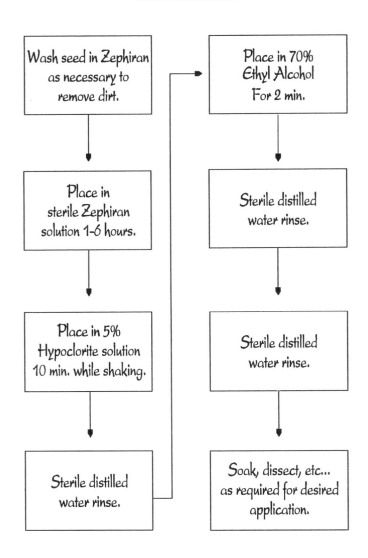

Wash seed in Zephiran as necessary to remove dirt.

Place in 70% Ethyl Alcohol For 2 min.

Place in sterile Zephiran solution 1-6 hours.

Sterile distilled water rinse.

Place in 5% Hypoclorite solution 10 min. while shaking.

Sterile distilled water rinse.

Sterile distilled water rinse.

Soak, dissect, etc... as required for desired application.

Other specialized techniques are sometimes utilized. The most common involve the production of haploid plants from reproductive structures, or various means of obtaining embryogenesis. Root or leaf tissue may be an appropriate explant source for some species. For special purposes, even protoplasts can be used to initiate cell or other cultures. All of these methods are described in later chapters.

Chapter 5

Embryogenesis

"Embryogenesis" has been used to describe several different processes utilized in plant biotechnology. The primary use of the term has been with reference to artificial culture of embryos obtained from seed. Another aspect of embryogenesis is the production of plants from embryos derived from somatic cells. The term has also been used to describe what might more properly be called embryo induction--the processes by which embryos are caused to develop in somatic cells.

Seed Embryos

Let us first consider the production of plants by culture of seed embryos. Contrary to some misconceptions, the seed is not the embryo, but it contains the embryo. The rest of the seed is composed of reserve nutrients and protective materials to support the embryo.

Figure 5.1 shows the major parts of embryos in a typical monocot (corn) seed on the right and a dicot (bean) seed on the left.

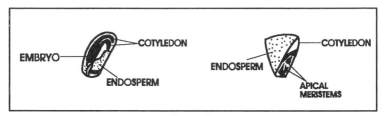

Figure 5.1

Commercially, there are a number of reasons for developing plants from embryo cultures. The use of such techniques provides a means for overcoming various problems:

1. Many embryos begin formation, but do not develop to maturity as a result of untimely death of the parent before the seeds have a chance to mature. In other cases, poor environmental conditions may prevent maturation of the seed.

2. Another use of embryo culture is to overcome inviability resulting from crosses between unrelated species. Quite often fertilization can be achieved between unrelated plants and the embryo will start to develop, but it may not proceed to maturity. Even when it does, germination of the seed is not likely to occur. Such problems can sometimes be overcome by removing the developing embryo from the seed before the seed reaches its full maturity. If the embryo can then be supplied with its complex nutrient needs, it may go ahead and develop.

3. Another reason for using embryo culture is to overcome prolonged seed dormancy which may extend to months or years. Often it can only be broken by a unique combination of physical and/or chemical factors. In the case of some parasitic plants, development will occur only if the host is present. Embryo culture can be used to overcome many of these problems. In many crosses of roses, for example, the long dormancy period results in excessive breeding time. If the embryo is removed from the rest of the seed at an early stage of development, immediate growth can be obtained, thus reducing by several months the time required to learn the results of the hybridization.

4. The use of embryo culture is an important research tool. One application is the production of plants from the seeds of species which must ordinarily be vegetatively propagated.

The banana is a common example of such a plant in which the seeds do not germinate well in nature.

Embryo Rescue

The artificial stimulation of plant development from seed embryos produced by the various conditions described above is sometimes referred to as "embryo rescue." This term is highly descriptive since extended dormancy or death of the embryo would likely occur without such artificial stimulation of growth.

Embryo Induction

Callus and other somatic tissues can sometimes be used as a source for the artificial production of diploid embryos.

In many cases wherein growth conditions are not ideal for shoot multiplication, the cells of the explant may spontaneously develop embryo-like structures called embryoids. In some cases, embryo formation can be artificially induced in cell cultures by manipulation of hormones, light intensity, and other growth conditions.

Following development of the somatic embryos, plantlet growth can be obtained by changing to appropriate stimulating hormones and achieving ideal light and temperature conditions.

If the cell culture is conducted in a manner that will prevent the adherence of cells to each other, millions of embryos can be produced in a very short time within only a few square feet of space!

"Artificial" Seed

A major logical and economically important extension of induced somatic embryogenesis is the production of artificial seed. The embryos can be coated with polymers to protect them during storage and distribution. If the right degree of water solubility is inherent in the coating, germination can often be achieved within a few hours of the time of planting. In the case of seed from plants which exhibit a dormancy requirement of several days or even several weeks, the benefits of such rapid germination are numerous. Of major importance is the potential for extension of the growth range of a particular plant. In many cases, reduction of the time required to achieve maturity by only one or two weeks may permit the growth of the plants in geographic areas several hundred miles removed from the normal limitations.

Figure 5.2 is a flowchart which summarizes a typical protocol for embryo induction from somatic tissues. As with any micropropagation procedure, strict attention to aseptic technique and a sterile growth environment is required.

Figure 5.2

Somatic Embryogenesis

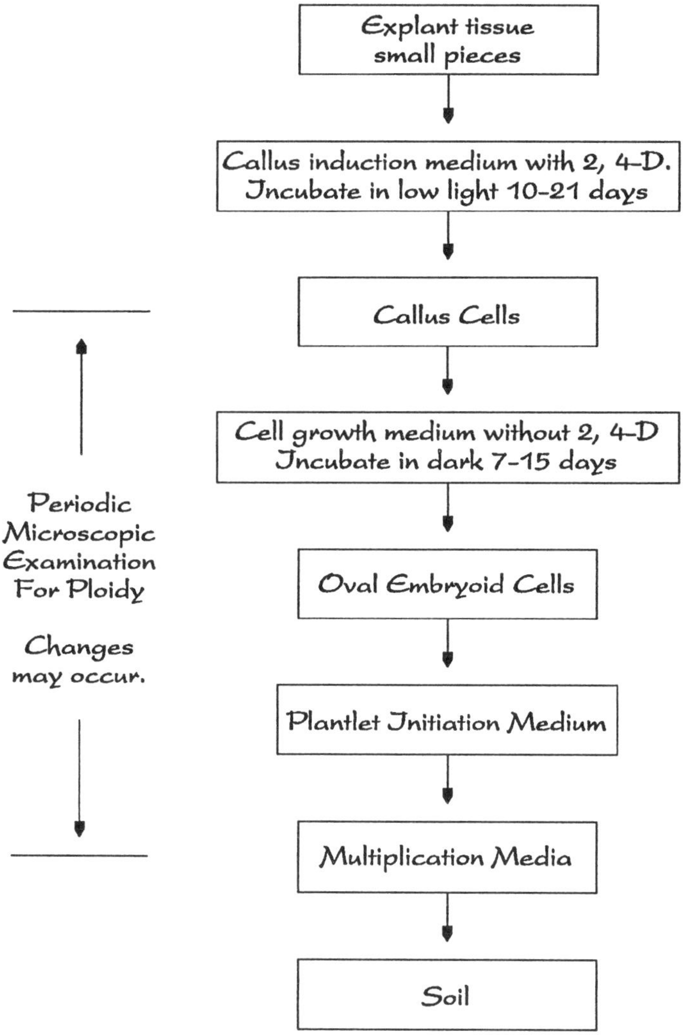

Did Aristotle's student, Theophrastus, around 300
B.C. have a vision of today?

"A plant has power of germination in
all its parts, for it has life in them
all, wherefore we should regard them
not for what they are but for what
they are becoming."

Chapter 6

Protoplast Fusion

This chapter is concerned with one of the most fascinating advanced new techniques in biology. Protoplast fusion is a procedure which engenders an overwhelming, majestic and almost reverent awe in those privileged to observe it. Our language has no words to describe the medley of emotion which almost transcends the scientific observations of the process. Most scientists have been well-taught not to become emotionally involved with their subjects. Can this ever be possible for one who is presiding at the creation of new form of life? That is the purpose and essence of protoplast fusion.

Protoplast Characteristics

Protoplasts are all the contents of a cell inside its wall. They are prepared by removing the cell wall without damaging the other cell components. Since the cell wall is made of materials like cellulose and pectin, it is fairly easy to obtain the protoplast by digestion of the wall with enzyme mixtures. When protoplasts come together soon after removal of the cell wall, two protoplasts will exhibit a strong tendency to fuse and merge the cytoplasm into a single cell structure. This fusion product will then regenerate a cell wall and begin to function as a normal plant cell.

Applications of Fusion

One of the major potential uses of protoplast fusion is the creation of new plant varieties by crossing unrelated species.

Hybridization of plants by conventional sexual crosses between related varieties has resulted in many economically valuable crops and ornamental plants. Two major problems exist with such techniques:

1. Time: often years are required to achieve a successful cross.

2. These hybrids are limited to closely related varieties within a single species.

Asexual fusion of cells from unrelated plants or different varieties holds great promise for the development of new species. The first phase of such a process involves fusing protoplasts from two unrelated plants. This is followed by organization of a cell wall and growth as a new cell. The second phase, which is the most difficult, involves differentiation into a plant and multiplication of it.

Figure 6.1 shows a fusion occurring. The original cells were callus from cantaloupe cotyledons and watermelon leaf mesophyll cells. The fusion was done as part of a 1993 science fair project by Colinda Roden, then a junior at Columbia-Brazoria High School, West Columbia, Texas.

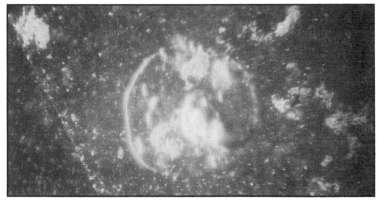

Figure 6.1

Protoplasts are obtained by digesting the cell wall with the enzyme cellulase. Polyethylene glycol or electric currents are used to aggregate the protoplasts to give them a greater opportunity to fuse at a faster than natural spontaneous rate. Since fusion occurs at random, many protoplasts of the same kind will fuse, producing homokaryons. These must be distinguished from the desired heterokaryons which result from the fusion of different species.

Figure 6.2 illustrates a common protocol for obtaining protoplast fusion.

Protoplast-Organelle Fusion

In many cases where hybridization cannot be obtained by fusion of unrelated protoplasts, some characteristics of one species can be transferred to another by the technique of protoplast-organelle fusion. Most commonly, this involves fusion of mitochondria (the respiratory organelles of a cell) to alter metabolism processes, or fusion of plastids to change photosynthesis, and in the case of chromoplasts, color characteristics. Although plastids and mitochondria have their own DNA, it does not code for all the materials needed to achieve *in-vitro* multiplication of these organelles. They will often, however, maintain their characteristics if fused and incorporated into a different type of host cell.

Plastids

Plastids, the photosynthetic mechanism of a plant, have their own DNA and ribosomes. Electron micrographs show that the DNA is present in thin strands located in the stroma. The DNA apparently codes for some protein synthesis as well as portions

Figure 6.2

Protoplast Fusion

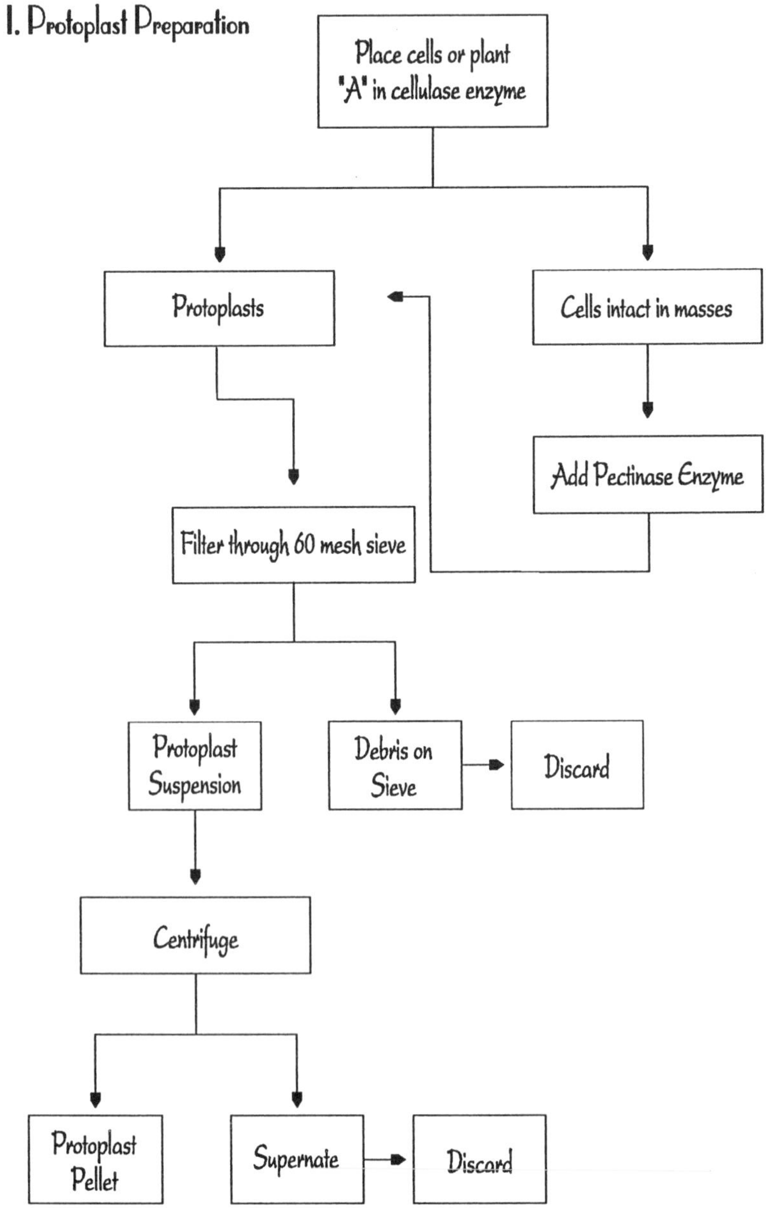

II. Fusion

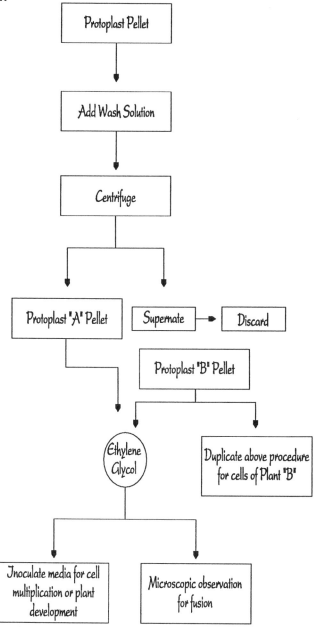

of the plastid reproductive process. However, reproduction is also controlled at least in part by nuclear DNA. It is interesting that some of the proteins coded by plastid DNA are also coded by the nuclear genome.

Most plant physiologists classify plastids into three categories. The chloroplasts are the most common type. These contain the chlorophyll, which is the major control of photosynthesis in most plants and gives leaves their green color. Another type of plastid is the chromoplast, which can be of a wide variety of different colors depending upon the photosynthetic pigment it contains. Some botanists feel that chromoplasts are actually a disorganized state of chloroplasts which lack chlorophyll, thus permitting the smaller quantities of other pigments to show. It is recognized that some of these other pigments are less efficient than chlorophyll in bringing about photosynthesis. This is not necessarily bad--it could well be a major means of adaptation for low-light plants. The third type of plastid is the leucoplast which lacks pigments. These usually do not have an internal system. Their function seems to be one of storage, particularly of starch.

Proplastids

The plastids can be reproduced from pre-existing ones or, in some cases, develop *de-novo* from undifferentiated structures apparently formed for that purpose. Such structures are usually referred to as proplastids. An area of continuing research revolves around a few reports that plastids can change their nature and become different types depending upon the metabolic needs of the plant at a given time.

Evolutionary Thoughts

A most interesting developmental theory about plant structure and function is that present plant cells originated as a result of an endosymbiotic relationship formed when free living photosynthetic cells consisting primarily of chloroplasts were incorporated into the structure of prokaryotic cells. A very readable discussion of the endosymbiotic hypothesis for the origin of prokaryotic plant cells is presented by Nadakavukaren and McCraken (1985). Many other reviews of this subject can provide a wealth of thought for private scientific philosophy or lively debate.

Figure 6.3 illustrates a typical procedure for protoplast-organelle fusion utilizing plastids from one species fused into the protoplasts of another. Similar protocols would be followed for fusion of other organelles such as mitochondria.

Microinjection

Another recently developed technique involves the direct injection of material from one cell into another. Variations of the process have been developed for use with everything from DNA fragments to whole cell organelles. A recently developed procedure is called electroporation. It involves the use of a strong pulsed electric current to drive the foreign material through pores of the cell membrane. Electroporation can often incorporate the desired molecules or microstructures into intact cells without the necessity for protoplast preparation. Special equipment to accomplish this is becoming commercially available.

Figure 6.3

Protoplast-Organelle Fusion

Prepare recipient and organelle donor protoplasts according to Figure 6.2

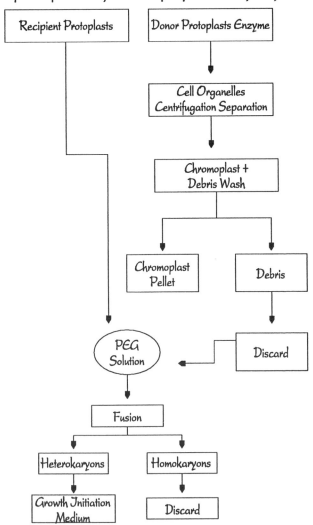

A Classroom Classic

All of us can recall a few truly memorable classroom moments--
never-to-be-forgotten learning experiences. A protoplast fusion
lab can be one of those. Few students (and teachers) can resist
the hypnotic vision of the merging of two life forms to initiate a
new one never before seen! Convenient kits are now available
to make it easy for instructors of college and advanced high
school courses to give their students this experience.

Over 2,000 years ago, Theophrastus described sex
in plants:

" 𝔚𝔦𝔱𝔥 𝔡𝔞𝔱𝔢𝔰, 𝔱𝔥𝔢 𝔪𝔞𝔩𝔢𝔰
𝔰𝔥𝔬𝔲𝔩𝔡 𝔟𝔢 𝔟𝔯𝔬𝔲𝔤𝔥𝔱 𝔱𝔬 𝔱𝔥𝔢 𝔣𝔢𝔪𝔞𝔩𝔢𝔰,
𝔣𝔬𝔯 𝔱𝔥𝔢 𝔪𝔞𝔩𝔢 𝔪𝔞𝔨𝔢𝔰 𝔱𝔥𝔢𝔪
𝔯𝔦𝔭𝔢𝔫 𝔞𝔫𝔡 𝔭𝔢𝔯𝔰𝔦𝔰𝔱."

WHAT WOULD HE THINK OF THE NEXT CHAPTER?

Chapter 7

Haploid Plant Production

Haploid plants are those which have only one member of each pair of chromosomes. With normal sexual reproduction of plants, the diploid, or in some cases, polyploid chromosome number remains constant from one generation to the next. This occurs as a result of cross fertilization in which one member of each pair of chromosomes comes from each parent. As the gametes are formed, the process of meiosis results in the reduction of the chromosome number to one half of each pair. Normal sexual reproductive processes give opportunity for variations induced from different parents.

In many instances, it is beneficial to stabilize the genome of a plant. This is particularly true when a single plant might show unusually valuable characteristics as a result of mutation or some unique combination of genes from a parent. If the required characteristic can be obtained from one member of a pair of alleles, plants could vegetatively be produced from it. All would have characteristics of the single desirable parent. A haploid plant, of course, will often be sterile. If this is a matter of importance, haploidy can be changed by the use of chemical treatments such as colchicine or other methods to multiply ploidy.

Methodology

Haploid plants can be produced in a number of ways. One of those is dependent on abnormal embryo development, which can sometimes result in haploid twin seedlings. This method can be controlled only with great difficulty and has not proven to be a practical approach. The two methods most commonly used involve production by means of tissue culture from either immature pollen or from cultured ovules if a female is to be the parent. The first production of haploid plants from pollen or anther culture was achieved by Guha and Maheshwari in India in 1964. Their work was studied and repeated. Nitsch, in 1967, developed the first haploid plant from isolated anthers of *Nicotiana* species. Since that time, numerous plants have been produced by either anther or ovary culture. The process has not succeeded with all plants. There seem to be major differences related to genus and species. Woody plants in general have been less successfully cultivated in this manner than have the herbaceous plants. Kott and Kasha (1985) gave a comprehensive review of haploid plant production, history, and technique.

A practical application of the use of haploid plants is described in the section (Chapter 11) on new foods.

Androgenesis

By far the most common method of producing haploid plants has been production from anthers. Individual pollen grains are technically difficult to work with. There are two means of obtaining haploid plants from anther culture. One involves direct embryo development from the microspores of the immature anther. The other depends upon the production of callus from the anther tissue. The latter method has some limitations in that diploid anther structural cells could give rise to the callus instead of the haploid microspores producing it. The source of the callus would be difficult to determine initially without sacrificing some for staining procedures.

A flow chart of typical steps in obtaining androgenic plant development is shown in Figure 7.1.

Figure 7.1

Androgenesis

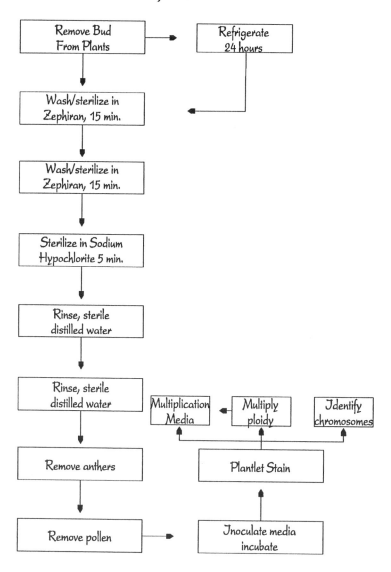

A major problem in any type of plant tissue culture is obtaining sterile starting materials and maintaining sterility throughout the period of culture. This initial problem is usually eliminated with minor care in technique. The most successful anther cultures are obtained from anthers which are removed from the flower buds before they have opened. Theoretically at least, there should be no bacteria or mold spores within the unopened bud. It is possible to sterilize anthers from open flowers, but the success rate of development in the resulting cells is much lower than that obtained from immature buds. Most workers have the greatest success by obtaining the anther at the stage in which the pollen grains have not undergone their first division.

The bud stage shown in Figure 7.2 is most likely to yield pollen at the ideal stage of maturity to obtain androgenesis.

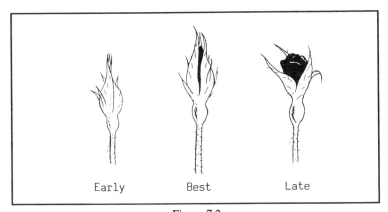

Figure 7.2

Another problem in anther culture is that quite often haploid plantlets will arise from individual pollen grains, and diploid or even polyploid plantlets begin at the same time from the somatic anther cells. In this type of situation, the faster growing diploids or polyploids will often crowd out the usually much more slowly developing haploid.

Ovagenesis

The use of the ovary or ovule in producing haploid plants has been much less successful than the use of androgenesis. This may be the result simply of lack of experience among workers in the field.

The anthers are usually much easier to obtain and to manipulate in the laboratory process. Quite often they can be shaken loose from the flower, whereas at least some degree of dissection may be necessary to get to the proper ovarian material. The results are generally identical. There is more tendency for ovary material to produce diploid or even polyploid plants than result from androgenesis. On the other hand, some sex-linked traits are transmitted only on the X chromosome.

Even so, one of the major advantages of ovule culture is often the obtaining of more genetically stable plants. There seems to be less tendency for undesirable characteristics to occur. In some cases, flower colors different from those of the normal diploid plant will result. There is also evidence in the literature that the success rate with ovule culture may be more species dependent than that for anther culture.

Problems in Haploid Production

Some of the difficulties associated with haploid plant production have been alluded to above. There are many others. One of those is the time-consuming nature of the process. This makes it economically unreasonable for many types of research. Many of the other factors can result in consumption of large segments of laboratory time which relates directly to the economics of the process. A major problem is the variation in nutrient requirement for different species. Tremendous numbers of nutrient variations often have to be made before androgenesis is achieved. Media selection can be complicated by inherent variations in requirements for different physical factors such as daylength, irradience, light to dark temperature cycles, etc.

It is not unusual for a plantlet which begins growth as haploid to spontaneously multiply its ploidy to diploid or tetraploid. In some cases this may be desirable; in others, it is not. Another problem experienced, particularly with haploid production in the cereal grains, is an unusually high incidence of regeneration of albinos from anther cultures.

Interferences

Several workers have found that there is a small quantity of abscisic acid present in the anther structure of many flowers. The acid will often prevent generation of vegetative material. Some success has been achieved in overcoming such problems by utilizing activated charcoal to absorb the acid. This, however, introduces other problems in that the charcoal may also absorb trace elements required for plantlet growth.

Continued research is gradually overcoming many of these problems. In some cases the answers turn out to be quite simple. For example, simply chilling the flower to a temperature of 10° to 15° C will often cause embryogenesis to occur when it would not if the flower had remained at an ambient temperature.

"Supermales"

Despite the problems, research continues in the production of haploid plants. They furnish a valuable tool for increasing our knowledge in many areas of plant reproduction and physiology. In some cases, obvious economic advantages have justified the research investment. A good example of this is in the production of asparagus where male plants which are designated XY have a much faster growth rate and yield the edible portions earlier in the season than do the XX female plants. Normal sexual crosses would result in a fifty-fifty ratio of XX and XY plants. Hondelmann and Wilberg (1973) showed that doubling the Y chromosome of the haploid male plant resulted in what they termed a "supermale" YY which could be vegetatively propagated. This plant was much larger and much faster growing than the normal XY, female XX, or the haploid single Y.

Simple androgenesis experiments can be carried out in most school laboratories, although it is unlikely that a haploid plant can be fully grown without some special equipment and supplies. SYNTHEPHYTES makes a laboratory kit, "ANDROGENESIS," which contains all materials required to obtain a successful anther culture.

It is
characteristic of
science and
progress that they
continually open
new fields to our
vision.
--Pasteur

Chapter 8

Chemicals From Plants

When asked the source of atmospheric oxygen, most people quickly think of plants, but our chemical dependence on them extends far beyond that basic need. The green kingdom is also the source of hundreds of useful chemicals for both drug and industrial purposes. These are usually obtained from plants which can be produced by normal mass agricultural practices. However, more and more potentially valuable materials are being discovered in species which are not adaptable to large scale agriculture. This group of compounds includes many fragrances, possibly useful drugs, and other phytochemicals.

Replaceable Resources

As early as 1983, Thorpe *et al* listed a large number of chemical compounds obtained in commercial quantities from plant sources. From table sugar to rubber, from maple syrup to cooking oils and fine waxes--with a moment's consideration one can think of many large-quantity plant items used daily. However, most people (and perhaps their physicians as well) are shocked to discover that over two-thirds of our present prescription drugs are based on plant material.

Dougall (1979) reviewed the literature relative to metabolite production *in-vitro* and found that cell cultures of 16 plant species produced as much or more of the chemical desired than was found in the whole plant. Many more have been found since that early study.

One of the new technologies utilizes cell cultures to produce a valuable chemical which in the past had to be extracted from a field-grown plant.

Modern chemical analytic techniques are revealing an almost uncountable number of previously unknown compounds present in plants. Many of these substances have potential medical value. Unfortunately, they are a frustrating as well as intriguing subject of drug research because many are available only in ultramicro quantities from rare or exotic plants which cannot be mass produced. The frustration and concern of researchers is compounded by the fact that many of the source species are near extinction, or are tropical varieties threatened by mass destruction of the rainforest areas.

Many of the drug chemicals isolated from plants are well known. They include digitalis, heroin derivatives, quinine, and even newer ones such as the anti-cancer compounds, vincaleukoblastine, and Taxol. Some of these are now being produced by cell culture as well as traditional plant extraction.

In-vitro Production

Any meaningful approach to phytochemical production *in-vitro* must distinguish between primary and secondary metabolites. Primary metabolites are usually defined as those compounds normally produced by the plant in accomplishing its regular life processes of growth, reproduction, and other basic functions. Secondary metabolites are often thought of as substances produced by the plant in adapting to its environment or in defense against attack by pathogens and predators. Generally, the primary metabolites are produced *in-vitro* in the same manner and relative quantities as found *in-vivo*. The secondary metabolites, particularly if they are adaptive in nature, may be produced in much smaller yield or may not be produced at all.

Elicit And Inhibit

Considerable attention is being given today to the use of elicitors which will initiate the desired biosynthetic pathways *in-vitro*. For example, if the biosynthetic precursor of the desired product is known, but is not present in a single cell type culture, it can sometimes be provided exogenously in the nutrient medium. Similarly, compounds which might be inhibitory for the metabolic pathway might be blocked or removed from the medium. The ability to adjust these nutrient factors to optimum levels for metabolite production is one of the major advantages of *in-vitro* production.

Cell cultures for phytochemical production can be carried out with adaptation of a number of standard techniques. Regardless of which one is used, a cell culture is obviously necessary to begin the process. Figure 8.1 illustrates a fairly standard approach to obtaining the desired cell line.

Following establishment of the cell culture, the next step is to determine optimum physical and chemical conditions for maximum metabolite production. This procedure often involves months of testing and evaluation. In many cases, not only is the quantity of a substance important, but also the ratio of nutrients or hormones to each other may be an equally major consideration.

Two Roads To Travel

Depending upon quantities of the metabolite required, along with factors related to culture processes, two approaches have been used for metabolite production. The "batch" process could be compared to the growth of bacteria in a single container. The

Figure 8.1

Establish Starting Culture

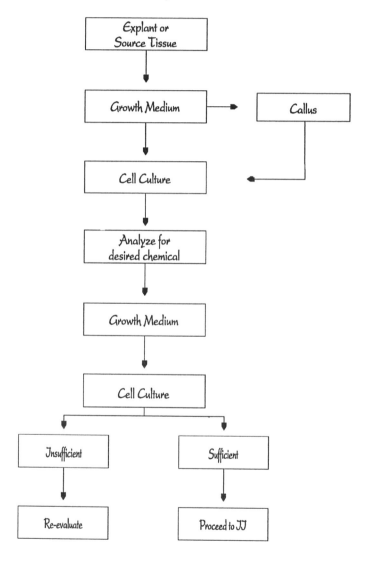

cell growth will follow a pattern similar to that of a classic bacterial growth curve. When the optimum concentration of the desired chemical is reached, the culture is terminated and the

Figure 8.2

Production Processes

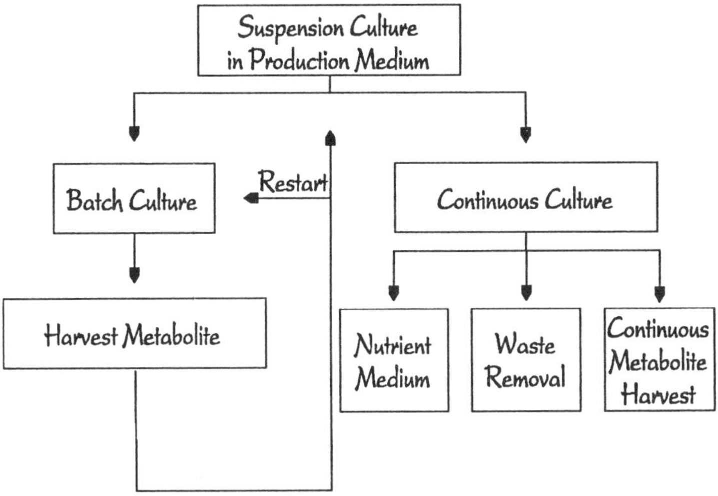

product extracted. Figure 8.2 shows a general protocol comparing batch production with continuous processes described below.

In the continuous culture processes, the cells are maintained in the maximum production characteristic by constant addition of fresh medium, removal of possibly toxic materials, and continuous extraction of the desired metabolite. An advantage of a continuous system is that it provides for steady growth and production rates. A serious disadvantage is that because of the constant cell multiplication occurring, chances for undesirable mutations increase. A program of vigilant monitoring for new cell line development is an integral part of the process.

Chapter 9

New Species And Varieties Development

One of the most fascinating areas of plant biotechnology is that of new species development. The reasons for creation of new species and varieties are still the classic ones taught in every high school Biology I course:

1. To extend geographic and climatic ranges of a species.

2. To select varieties which have natural disease resistance.

3. To select varieties which can adapt to unusual physical conditions.

4. To develop new plants which have more useful characteristics such as increased food value, beauty, etc.

5. To develop varieties or cultivars which will meet rigid size characteristics.

Old Ways And New Ways

Classical development of new varieties has been based upon careful selective pollination for cross breeding followed by conventional growth. With these methods, many years may be required to obtain the plant desired. Another method frequently utilized is that of careful observation for and selection of naturally occurring mutants. Even after the grower found a mutation, he still faced many years of breeding and development to obtain sufficient stock for commercial propagation. With both types of breeding stock, the value and genetic stability of the plant often could not be established without several generations of inbreeding.   Frequently this process resulted in the development of undesirable recessive traits. Utilization of new technology has greatly speeded the process, although the basic principles remain the same.

In addition to adaptations of the old processes, modern plant biotechnology employs the new tool of genetic engineering to meet specific goals. The process basically involves obtaining a gene for the desired characteristic from one species of plant and inserting it into the genome of another which will be improved with the new characteristic. The methods used are described in the next chapter.

Other approaches include use of the techniques of embryogenesis, protoplast fusion, and development of haploids previously described in detail. All of these are adaptable to establishment of new varieties.

Mutants By The Millions

When it comes to selection of naturally occurring variants, the biotechnologist has several distinct advantages over his classical horticultural predecessor. One of these is in playing the numbers game. A natural mutation rate occurs in cell cultures as well as *in-vivo*.   In fact, some biologists think the natural rate is somewhat increased under artificial conditions.   Thus, in a laboratory culture, the biotechnologist has literally hundreds of millions of cells to produce variations. Compare the ease of detection and selection here with plodding through thousands of acres searching for a unique plant!

Figure 9.1 shows the steps in this procedure. It is essentially the process used by DNAP in producing a new tomato (Evans *et al.*, 1988).

Another advantage is that of time. Once the variant has been found, cell multiplication and clonal propagation make an unlimited number of plants available within a few months, whereas the classical procedures might require years.

The process of mutant selection *in vitro* can be speeded up even more by the use of mutagenic agents. Control and application of chemical mutagens or radiation is much more easily accomplished than on a greenhouse or field scale.

Adaptations

If one does not wish to pursue mutations, selections of variants which naturally adapt to unusual conditions is a process which can be accomplished with relative ease. Figure 9.2 shows how selection might be made to obtain a strain which is tolerant of alkaline growth conditions. The cell culture or multiple clones are started in the usual way. They are initially grown on a nutrient which contains slightly more than the normal maximum alkalinity. Plantlets which survive and grow will be taken from this nutrient and placed on one at an even higher pH. The process of selection in this manner would be continued until a few survivors could tolerate the alkaline condition. These would then be multiplied *in-vitro* to provide sufficient material for field testing on a large scale. A selection process of this type could conceivably reduce by years the time required to find a variant tolerant of alkaline soil conditions.

Another method sometimes used for selections of adapted strains is illustrated in Figure 9.3. It shows a variation of the bacterial gradient plate method of Lederberg used in the 1950's for studies

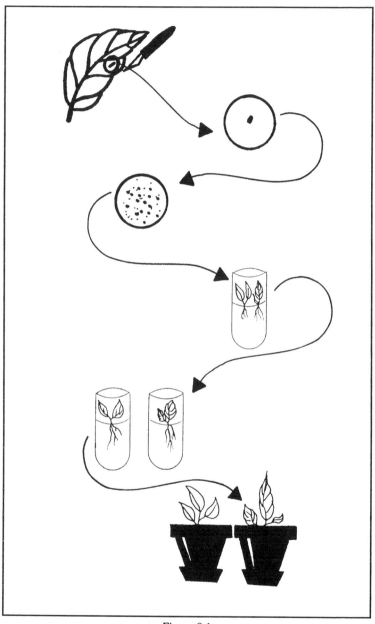

Figure 9.1

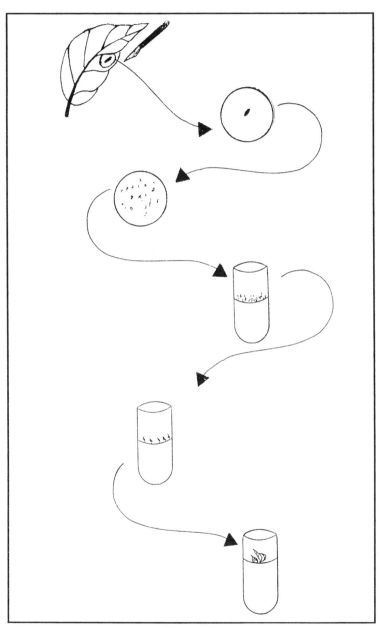

Figure 9.2

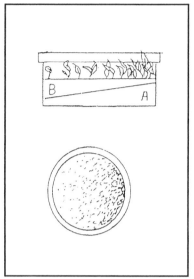

Figure 9.3

of antibiotic tolerance in bacteria. This process works best with adaptation to compounds that have limited water solubility. As in the previous technique, tolerant variants would be selected and subcultured to gradually increasing higher concentrations of the test compound.

The processes described above along with new ones currently being developed hold great promise of providing man with hundreds of new plants species within the next few years.

Chapter 10

Recombinant DNA

"Recombinant DNA" is a term which often evokes strong emotions of fear, mistrust, and animosity. Science fiction has reinforced and magnified the imprecise understanding many people have.

One definition of genetic recombination involves movement of a gene from its normal position on a chromosome to another position, or in some cases such as crossing-over, it may involve movement to a different chromosome. The idea that recombinant DNA is a new part of biotechnology is erroneous. Actually, it is a natural process which occurs frequently. It was first described by Lederberg and Pattem in 1947. Another form of naturally occurring recombination was described in 1950 by Barbara McClintock. Her contribution, which involves a theory of random gene movement, was not widely accepted for many years; in fact it was twenty-three years later before she was finally recognized with a Nobel Prize in 1983. Our language is not static. The term recombination today is almost always associated with DNA; in the form of recombinant DNA, it has become part of the terminology of the new biotechnology. The natural process of recombination, which has occurred spontaneously since life first began, is almost totally overlooked because we think of the idea as being a modern advancement. Since we are concerned in this book with man's manipulation of natural processes, we shall confine further discussion of recombinant DNA to its modern connotation involving artificial manipulation. An excellent discussion of natural recombination is presented by Raven and Johnson (1989).

A more recent term coming into widespread use is "metabolic engineering." It was coined by Bailey (1991) to describe the use of recombinant DNA to change regulation of metabolic pathways in contrast to the original uses of coding for the production of a protein to be isolated.

Basically the techniques of recombinant DNA involve cutting of a DNA structure by an enzyme, the insertion of DNA obtained from another source, and the rejoining of the DNA. The new DNA so formed will include the "foreign" DNA spliced into the original structure. Different techniques are used to accomplish this process.

Plasmids

Plasmids are circular structures of DNA found outside the primary chromosomal DNA in many bacteria. The structure is illustrated diagrammatically in Figure 10.1. Since plasmids can be easily separated from the other contents of a bacterial cell, they make an ideal medium for an artificial DNA recombination. A further advantage of the use of plasmids is that they can

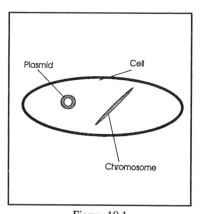

Figure 10.1

be easily cut enzymatically and spliced back together after incorporating the foreign genetic material. Once the new gene has been spliced into the plasmid it will be reproduced along with the other genes as the plasmid DNA is replicated. A further advantage is that the plasmid will be taken up when placed in contact with whole bacterial cells. This is the simple means of getting the recombinate plasmid into a new intact functioning cell. Once they have been incorporated, the cell will carry out the functions programmed by the recombined DNA of the plasmids. This, for example, is how a human gene can be incorporated into bacteria causing them to produce the enzyme or other chemical substance coded by the gene. A practical application of such techniques is the production of human growth hormone by bacteria which have been modified by the addition of the human gene for that substance.

Similar techniques are utilized in modern plant biotechnology. At this point you might be thinking, "So the gene can be transferred to bacteria and reproduced hundreds of millions of times. So what does that have to do with getting it into a plant?"

That was a good question. One part of the answer is that it provides a quick and easy means for making countless copies of the desired gene. The second part of the answer involves transferring the replicated gene into a plant to achieve the desired characteristics.

This brings us to the term "genetic engineering." Is that term frightening? It shouldn't be. Quite simply it is used to describe the artificial incorporation of a foreign gene into a host.

Did that definition blow away some esoteric concepts of modern biotechnology? Sorry 'bout that. We are dealing here with practicality and tedious hard work. It has fantastic potential, but it is not, as some would like to believe, an incomprehensible fearsome incarnation of science fiction. Granted, if misused the technology could conceivably produce some kind of monster like those popular in Hollywood. Recombinant DNA might be compared to a gun. Guns do not kill. People misusing them do. Those who oppose the use of recombinant DNA because it has possible potential for bad as well as good, might be asked if they

would be willing to give up driving a car because people get
killed in automobile accidents. They also got killed by falling
off horses. With that digression out of the way, it is time to
proceed with the how to.

One of the major problems of genetic engineering of plants has
been that of finding a method to get a desired new gene into the
plant. Unlike bacteria, plants do not have plasmids which make
the transfer relatively easy. Until recently, the most widely used
method has employed bacteria or viruses as vectors to transfer
the desired gene. The most satisfactory method to date involves
using the bacterium
Agrobacterium tumifaciens
which infects many broad leaf
species of plants.

The plasmid DNA of the
infecting *A. tumifaciens* will
combine itself with the plant
DNA. Thus, any gene which
is located in the bacteria
plasmid can become part of
the plant cell genome
following infection. Cells
which incorporate the desired
gene can then be removed
and used as starting material
for normal micropropagation
processes. The resulting
plants will contain the new
gene. Figure 10.2 illustrates
one common procedure of
preparing an *A. tumifaciens*
plasmid for transferring a
foreign gene into a plant.

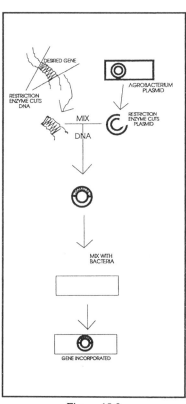

Figure 10.2

Other methods are being sought for plants which are not infected by *A. tumifaciens*. This includes a vast number of economically important species such as the grasses and cereal grains. For these, transfer of genes by means of a virus or viroid vector is a possibility. The procedure would be similar to that employed with *A. tumifaciens*. Incorporate the desired gene into the viral DNA or viroid RNA. When such organisms infect a host plant, they will transfer the new DNA code, making it an intergral part of the plant's genome.

One of the advantages of these methods is the fact that plant pathogens - bacterial, virus, and viroid are highly host specific. Not only are they limited in their capabilities of infecting a broad range of plants, but also there is the additional advantage that they are totally incapable of infecting any animal cells.

Direct Methods

All plants are not susceptible to infection by the microbes just described. For these plants other methods must be used. They all involve some type of direct injection of the new DNA into a plant cell.

One such method is called electroporation. The plant cells to be modified are suspended in a liquid containing the new DNA which the scientist wishes to transfer into the plant. The mixture is then subjected to a momentary shock with a strong electric current. The shock causes pores in the plant cell membrane to open wider than normal. As they open, the DNA surrounding them moves into the cell. It can then incorporate into the plant's original DNA to provide the code for the desired new characteristic.

Another method is referred to by a variety of names including microballistic injection, particle guns, or gene guns. Regardless of the name used, it involves a "shotgun" approach. The DNA to be incorporated into the plant cell is coated on microscopic pieces of gold, tungsten or zinc. These particles are then blasted under pressure into the cell. If all goes well, the desired DNA will be incorporated into the original DNA.

Microinjection is another method. It is exactly what its name sounds like. A microscopic needle is inserted into a single cell. The new DNA is injected through it into the region of the nucleus. Again, as with the other methods, the new DNA may combine with the original, thus adding the desired characteristic. The disadvantages of microinjection include the simple physical difficulties of carrying out the process and the fact that it can be used on only a single cell at one time. Cells are also more likely to be injured than they are with the methods described above.

The idea that transgenic plants might produce human or lower animal mutations with their genetically engineered genes is no more likely than that any normal plant gene will cause a mutation in an animal cell. There is no more to fear from these than there is to fear from the genes of any of the plant foods we already consume. The new gene simply becomes a part of the plant's genetic code.

Part III

The Products

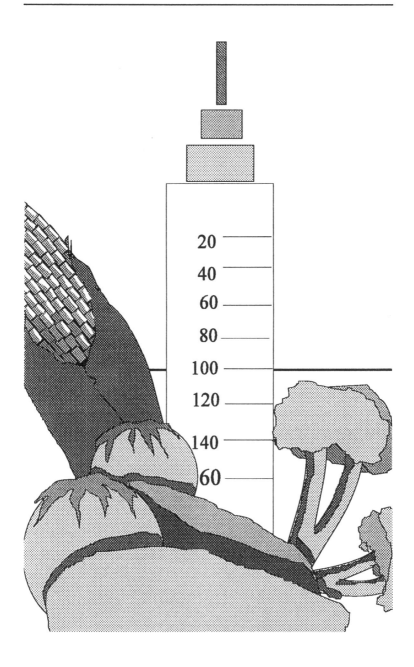

The Products

Since earliest recorded history man has attempted to make improvements to the bounty which nature supplied. History is replete with instances of efforts to enhance desirable characteristics of both plant and animal. Long before Gregor Mendel discovered the basic principles of genetics, man made practical applications for the improvement and selection of the most desirable plants. As early as the 17th century, rose fanciers were hand pollinating to obtain new hybrid varieties of the plants originally brought to Europe from China. Farmers might carefully save beans from the most proflic bush for planting the next year, or save the ear of corn which was produced by a plant with more desirable characteristics than the others in his field. Later, as the principles of genetics became known, agriculturists practiced careful selective cross breeding to bring about more desirable varieties.

To select from and improve on the natural plants is nothing new. The difference between the old ways of hybridization and the new processes based on genetic engineering or other new technologies is primarily one of time. Today's method also has the advantage of making it possible to select one or a few specific traits rather than being forced to adopt a random chance approach. In some respects modern biotechnology procedures might be compared to driving a new automobile. It gets you there faster, but it still runs on four wheels just as Henry Ford's Model T did. Modern technology is giving us a desired result in much less time than the years or even generations required by traditional cross breeding and hybridization methods.

The rest of this section will explore the fantastic and awesome green products now becoming available.

Chapter 11

Improved And New Foods

Cactus Chili And A Chaser

Is this any way to start a chapter about great new foods? Cactus? Perhaps, not to many North Americans, but to millions of others throughout the world, dietary use of cactus has a long, distinguished, and delicious history. A type which is receiving extensive attention is the kind commonly called prickly pear (genus *Opuntia*). Many varieties of *Opuntia* species are found widely distributed in arid regions of the world. The prickly pear serves as cattle forage in many areas, particularly when drought conditions become severe and the native grasses become scarce. Ranchers use propane torches to burn the spines from the cactus pads so that cattle can eat them without injury to their mouths. The prickly pear is an important part of the diet of many in northern Mexico and is considered a delicious food in Israel and many other parts of the world.

Both the reproductive and vegetative structures are useful. The large flat green pads which make up most of the body of the cactus are used to make a food known as "nopalitos." The pear shaped fruits which form on the pads are known as "tunas." Recipes are becoming more common as many become acquainted with the prickly pear. The tunas are used in making jelly, candy, and a variety of other dishes. The pads are used for the popular Mexican nopalitos which can be prepared in a variety of ways.

Author Elizabeth Schneider, in her book <u>Uncommon Fruits And Vegetables</u>, describes nopalitos as being "soft, but crunchy with the flavor of green pepper, string beans, and asparagus, all touched with a citric edge and the slipperiness of okra." The biggest problem with prickly pear is that the plant is well equipped by Mother Nature to defend itself against all who have designs on it for food whether cattle, wildlife, or human. To avoid injury, the spines must be removed by burning before it can be a suitable forage crop; similarly, the major problem in preparing nopalitos for human consumption is the tedious by-hand process of cutting off the spines.

Biotechnology comes to the rescue. Dr. Peter Felker and his group at Texas A & I University are developing several varieties of spineless prickly pear. These apparently originally appeared as an unusual natural mutation in Mexico. Felker has succeeded in micropropagating several of the spineless varieties to obtain sufficient material for field testing with respect to drought and temperature hardiness, taste, and other characteristics. Over 80 cultivars are being studied. The lack of spines means that these varieties could be consumed directly as cattle forage and would eliminate the most costly and time consuming steps in preparing them for human consumption. By employing laboratory cloning of the desirable varieties, Felker is reducing by years the time that would be required to bring them to commercial numbers for widespread planting.

Figure 11.1 is a picture or a common prickly pear; Figure 11.2 shows a spineless variety.

Early research provided some ideas that the prickly pear has healthful properties. There was even a patent issued in 1937 (Gruwell, *et al.*) with respect to the extraction of chemical substances which could be used in treating diabetes mellitus. The value of the anti-diabetic substances from prickly pear has not been scientifically established, but there is no question that the plant provides a very low fat, high fiber, high vitamin C content for the diet. Its other attributes include low sodium and

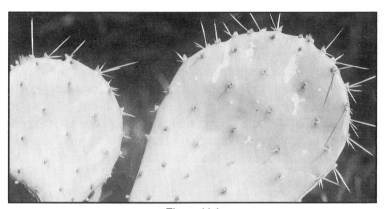

Figure 11.1

Figure 11.2

relatively high protein content.

Both nopalitos and tunas are becoming more widely available in supermarkets.   Here are some recipes you might want to experiment with. They have been supplied by Bruce Kraatz of

Riviera, Texas. When you have finished the main course, top it off with Mrs. Margarita Hinojosa's prickly pear pie. Remember that only the original spiny varieties are available at the time of printing this book. If the spines are still present, use caution and cut them out before making any further preparations.

South Texas Chili

1 cup diced Nopalitos

2 lbs ground beef

1 can tomatoes

4 tsps chili powder

1 tsp camino

1 tsp paprika

1 large onion

salt and pepper to taste

Brown the ground beef in a deep skillet. Add the other ingredients, cover with water, and simmer for one hour. This is not considered a hot chili. The term "hot" is not applied in South Texas unless it makes the eyes water and the nose burn and run. You might want to proceed with caution if you are not used to this. On the other hand, If you find it too mild, a little fire can be created by the a addition of some chopped jalepenos or other peppers.

Cactus Casserole

6 small potatoes

1 cup cheddar cheese

1 small onion

2 eggs

1 cup diced Nopalitos

1/2 cup butter

This is for skinny people who are not concerned about fat and cholesterol in thier diet.

Slice potatoes and onion. Mix all ingredients together and pour into a buttered casserole dish. Bake at 350 degrees F. for 30 minutes.

The Chaser: Prickly Pear Punch

Squeeze 2 cups of juice from Tunas. Mix with 2 cups of orange juice and 1 cup of crushed pineapple. Chill off, add ice. Just before use add one 32 oz. bottle of Sprite, 7-up, or Sparkling water to serve from a punch bowl. This is a delicious, refreshing, and healthful drink. You are on your own for the addition of any other ingredients.

Cactus Jelly

12 Tunas

6 cups sugar

1 pkg. fruit pectin

Be sure all spines have been removed from the Tunas. Cut them into small pieces and cover with water and boil for a few minutes. Strain, mix the juice, pectin, and sugar. Cook for 3 minutes and pour into sterilized jars while still hot.

The Dessert: Prickly Pear Pie-9"

3 1/2 cups boiled, diced nopalitos

1 cup sugar

3 Tbsp. all purpose flour/cornstarch

3/4 tsp. cinnamon

1 Tbsp butter

Pastry for double crust pie

Boil nopalitos in 5 to 5 1/2 cups of water. Let them boil for 3 minutes. Rinse the nopalitos under running water and then drain for a few minutes. Measure 3 1/2 cups nopalitos in a mixing bowl. Add all other ingredients except butter. Mix well by hand and pour into a 9" pie shell, dot with butter and top with crust. Bake at 425 degrees for 40 to 45 minutes or until golden brown.

This recipe for nopalito pie was provided by Margarita C. Hinojosa of Benevides, Texas.

Sweet Potatoes And Social Problems

Sweet potatoes are thought to have originated in South America and made their way to Africa during the 16th century. They easily adapted to growth in many parts of Africa and have become a major part of the food supply in countries such as Kenya and Uganda. Actually, the world's largest producer of sweet potatoes is China.

Although sweet potatoes are a popular, needed, and economically important crop in Africa, their production is much lower per acre than it is in other countries. The major cause of this seems to be virus infections, particularly one known as the feathery mottle virus. Yields from infected crops may be no more than one-fourth to one-half of that normally expected.

In the mid 1980's Dr. Florence Wambugu in Kenya began studying biotechnology methods which might control the loss resulting from viruses. The research with conventional crop improvement processes resulted in frustrating failure time and again until 1991, when Dr. Wambugu received a grant from Monsanto Chemical Co. and the U. S. Agency for International Development. Since that time, she has worked closely with Monsanto scientists to study applicable recombinant DNA processes. At the present time she has succeeded in inserting a gene which will assist in producing virus resistance. Wambugu is scheduled to return to Kenya in 1994 to begin extensive field testing of the promising newly developed varieties. She is shown in Figure 11.3 with Monsanto geneticist Dr. Robert Horsch. Previous work done by Monsanto has shown that combining the coat protein of the virus into the sweet potato will greatly increase the level of virus resistance by the potato. Wambugu feels that since the transgenic protein is already present in the environment, there is less chance for resistance to develop by the virus. If this proves to be the case, the sweet potato supply in African countries can be rapidly increased.

Figure 11.3

In a 1993 interview by Karen Freeman of Genetic Engineering News, Dr. Wambugu expressed difficulty in understanding what seemed to her to be exaggerated concerns about safety in genetically engineered foods. She agreed that safety tests are essential, but indicated the major concern of her work was providing a means to give hungry children a future.

It is easy for an overfed population which leaps from one crash diet to another to forget that calorie control is not a problem in much of the world. Dr. Wambugu has pointed out on several occasions that the problem in African and many other developing nations is that of obtaining enough calories for basic life functions. This problem will undoubtedly become more severe if predicted increases in populations occur.

The hard facts of poverty, hunger, and disease in much of the world raise many social and ethical questions. As the new plant biotechnology expands from the laboratory to the field, these issues will have to be considered. One that arises quickly is that of a biotechnology company's social responsibility. Does a corporation have a social obligation? If so, how can that coexist with its responsibility to make a profit for the stockholders who risked their savings to form the company?

By sharing its expertise in this joint effort with the Kenyan scientist, the Monsanto Company has addressed the questions and set an example of corporate social responsibility.   Some detractors of the chemical industry might say that a company such as Monsanto with its world-wide operations and vast resources from a long established broad product line could well afford to participate in this venture. While that might all be true, the simple and pertinent point is that they didn't have to. They did it. Perhaps a more accurate conclusion to be drawn is that the privilege of leadership carries with it special responsibilities.

The fact that many other plant biotechnology companies have not yet participated in such ventures is no criticism of them. Remember that most of these companies are young, small, and even after several years of research and development are not yet making a profit for their investors. Even so, some such as DNA Plant Technologies Corporation is making efforts along this line. That company has provided some of its technology to scientists from other undeveloped countries.

Much governmental and international agency effort to assist third world agriculture has resulted in misguided attempts to incorporate western high-tech farming practices. It is destined to fail because the countries needing help are simply not to the point where these techniques can be utilized. Aside from a few commercial operations aimed at producing high volume export crops, most of the agriculture efforts in developing countries are in the nature of subsistence family farming.   Biotechnology techniques directed to crop improvement by gene transfer may be the immediate answer. Monsanto crop scientist Dr. Robert Fraley puts it this way: "We can give them advanced technology in a package they understand. Every farmer knows what to do with a seed."

Tomatoes

From fresh to soup, tomatoes constitute one of the most popular food crops. They also constitute one of the largest problem crops. Everybody knows what a good tomato tastes like. Fresh from the vine, it has a unique combination of taste, texture and tenderness that producers, whether commercial or home growers, strive to obtain. Unfortunately, unless you grow your own, what you want is not what you get. To understand the problem just think about the last tomatoes you bought in the supermarket, especially during the winter. Enough said.

That unsavory experience was the result of producers and processors yielding to inexorable qualities and demands of mass production and distribution. In order to reach the produce counter, the tomatoes had to be picked before fully ripe. They also were varieties carefully selected for shipping and maintenance traits rather than the more desirable aesthetic qualities. It is not that the growers and sellers want to have an inferior product; it is simply a matter of that product or none. High quality fully ripe tomatoes would be mush by the time they reached the supermarket.

A few years ago when recombinant DNA technology first started becoming practical, several agricultural technology companies looked at the problems and potential in the food market. The United States uses almost 3 billion pounds of tomatoes per year. Even more mind boggling, perhaps, is the fact that the Campbell Company alone sells over 300 million cans of tomato soup every year. Certainly a market of this size, which was also one with a demonstrated need for improvement, would be a logical candidate for new technology. The results of research by several companies are almost ready for the table.

Probably the first genetically engineered tomato that will reach the market is one known as the "Flavr Savr" produced by Calgene, Inc. of Davis, California. Calgene geneticists claim that the "Flavr Savr" can be left on the vine long enough to develop the final flavor characteristics the consumer desires while remaining firm enough for shipping and handling. They have accomplished this by use of a technique known as antisense DNA. In this process, they incorporated into the tomato a backward copy of the code for a major enzyme responsible for fruit softening. The antisense DNA produces messenger RNA which binds to and inactivates the normal mRNA that carries the code for the softening enzyme from the nucleus to the ribosomes. This reduces the ability of the tomato to produce the softening enzyme known as polygalacturonase now being commonly referred to as PG. Since the tomato produces a lower than normal amount of PG, it can remain on the vine longer before harvest. The extra time gives the tomato a chance to produce the desirable flavor and texture components.

Any logical thinking scientist would immediately recognize that limiting the production of a naturally occurring fruit softening enzyme in a plant presents no threat to the consumer. Unfortunately, Calgene's efforts to get "Flavr Savr" on the market have been slowed by misguided groups threatening to file lawsuits to prevent it or other genetically engineered foods from ever reaching the table. Although the FDA recently ruled that genetically engineered foods do not require testing beyond that for conventionally hybridized new varieties, Calgene has sought specific approval for the "Flavr Savr." This is simply an effort to head-off delaying interference from those opposed to any use of genetic engineering.

Other Approaches

In addition to Calgene's efforts, other companies are pursuing different processes for tomato improvement. The Monsanto Company has taken the approach of using a gene to slow the production of ethylene gas. Ethylene is produced by most plants as a control for the ripening process of fruits. Slowing this production will retard ripening to provide far more efficient harvesting and handling. Other researchers are isolating additional genes involved in the ripening and flavor production processes.

A. K. Handa of Purdue University, in using the antisense technique to slow fruit softening, produced (1992) a tomato with approximately ten per cent greater solids content than most present varieties. High solids content is a matter of prime importance and economic value to those producing tomato products such as ketchup and tomato paste. If tomato varieties which have the desirable taste and aroma qualities can be improved to give a higher product yield from the same number of fruits, the result will be a boon not only for the processor, but also ultimately for the consumer. The results of the tomato research are not limited to tomatoes alone. The know-how and technology developed in this research will have direct application to improvement of many other fruits.

A Two-Pronged Approach

The DNA Plant Technology Corporation (DNAP) has a unique approach to the tomato problem. It is interesting from both the business and technical standpoints. The many years of research, development, and safety testing required to bring a new genetically engineered recombinant DNA food to market has created the same problem for this company that it has for many others - cash flow during the early years. DNAP has adopted a two-level approach to the problem.

The first level involves producing and marketing new improved vegetables that have been produced by biological processes which do not require regulatory agency approval. Essentially what they are doing is applying some of the methodology discussed in earlier chapters as advanced breeding techniques. One of the first results of this is a tomato which they have trademarked VineSweet™. This tomato was obtained by using the cell selection procedure similar to that described in chapter 9. The result of the speeded-up natural breeding and selection process is a tomato which has an extended shelf life of about 14 days compared with the average of seven days. The longer shelf life means that the grower can wait longer to harvest the tomato, thereby giving it a chance to develop fuller flavor on the vine. Test marketing has resulted in widespread favorable consumer reaction. Production and marketing of the VineSweet tomato along with other premium quality fruits and vegetables is being handled by Fresh World Corporation, a new company set up by DNAP in cooperation with the E.I. DuPont Company. Each owns fifty per cent of Fresh World.

The second level of approach involves the development of a tomato which will have even longer shelf life and, according to DNAP predictions, even more desirable qualities. It is being produced by recombinant DNA technology. DNAP refers to the process used as Transwitch™. The methods utilized are quite different from anti-sense DNA technology. The Transwitch™ process involves inserting a naturally occurring gene which will cause the plant to produce less ethylene than normal as it approaches maturity. Less ethylene means longer retention of the fruit on the vine and the opportunity for it to develop full taste characteristics. The tomatoes' qualities will permit them to be vine ripened before harvest.

DNAP claims that its patented Transwitch™ technology is several times more efficient than anti-sense procedures. Time will be the ultimate judge of their claim. One thing which is certain is that the DNAP technology provides another example of how a procedure developed for one product can be expanded for use in others. The Transwitch™ procedure was originally discovered in research of almost a basic nature dealing with petunias. DNAP's patent on the use of their process with petunias was expanded by the patent office in 1993 to cover many other plants. Another certain fact is that the DNAP process gives it a definite competitive advantage over other companies which are using the anti-sense procedures and at this time are embroiled in patent disputes over it.

Although the research is in an earlier stage than that with the tomatoes, DNAP is planning to use the Transwitch™ technology for improvement of a number of other crops including sugar peas, multi-colored peppers, Canola oil seeds, and various tropical fruits.

Grain Crops

The genetics of the major grain crops such as wheat, corn, and sorghum have been the subject of agricultural tinkering for generations. Selection for more prolific production probably extends as far back as the ancient civilizations of the Incas and Mayans. Building upon this early foundation, later cross breeding and hybridization efforts were directed primarily toward improving yields and making the grains more adaptable to modern mechanical farming. While achieving those goals, the breeding techniques lost some natural genes for disease resistance and other desirable characteristics. We pay the price today with massive use of insecticides and other chemicals to enable the plants to survive to provide their bounteous harvest.

Since the grains make up such a large base of both direct and indirect food supply for the whole world, improvement of their nutritional characteristics is a worthy goal. Incorporation of genes for vital materials such as certain vitamins and amino acids can go far toward relieving malnutrition among millions of persons, particularly those in the third world. Improved grain products are becoming available.

The amino acid known as lysine is essential in human nutrition. Unfortunately it is not produced in most grains, or for that matter, in many other popular food crops. A notable accomplishment in plant biotechnology has been the incorporation into corn of a gene which causes it to produce lysine. Thus corn, which provides a large portion of the basic food in many countries, becomes more nutritionally adequate.

Eliminate The Undesirable

Not everything that is present in the foods we eat is necessarily good for us. For example, two compounds naturally produced by many grains might be considered what **Consumers Research Magazine** in March and April 1991 referred to as "antinutrients." These compounds, known as oxylates and phytates, chemically bind to several mineral elements making them unavailable for absorption by the intestine. As a result, although the needed mineral is present in the food and would show up in a chemical analysis, it is nutritionally useless. This is another place were antisense DNA might be useful. By incorporating the antisense RNA, the production of compounds such as the oxylates and phytates could be blocked, thus making the natural mineral content of the food actually available.

A Look Ahead

Prediction, even scientific prediction, is often a dangerous gamble. Depending on the outcome, the predictor becomes a hero or a bum. In the case of plant biotechnology, we foresee many new foods undreamed of today.

A new type of natural sweet protein known as curculin has recently been described by Yamashita et al (1991). This protein and related ones have taste modifying characteristics. Once the genes responsible for making them have been identified, the proteins could be incorporated into various plants to make some of those vegetables really good instead of just good for you!

Some plants produce natural proteins which help protect them from freeze damage. Fruits and vegetables containing such anti-freeze proteins can be frozen and thawed without becoming mushy because of ice crystal damage.

In addition to the use of recombinant DNA, techniques of protoplast fusion and embryo rescue will provide the means for making many hybrids that would be impossible by traditional cross-breeding. Less undesirable variants of present foods is also coming. An early possibility is a potato with a higher than average solid content that translates into less oil absorption when fried. Combine that with a new Canola oil with less saturated fat, and you can enjoy those French fries without worrying about your arteries.

The Spin-offs

Just as NASA's space research has brought new products and knowledge to many unrelated areas, the new plant biotechnology will ultimately result in a bounty of useful benefits. Many of these are already becoming apparent.

Some are in the realm of expansion of basic knowledge with benefits yet to be determined in the future. Others already begin to cross the boundary between basic and applied research. A good example of this is a change in attitude about the science of ethnobotany, which studies the relationships of plants to specific populations. A few years ago, this science was only a curiosity confined to a few ivory towers of academia. New technology has provided the tools, and the decimation of our rain forests for transient economic gain has spurred a mad rush to follow in the footsteps of early plant explorers such as those described in **Green Medicine** (Krieg, 1964). Perhaps Margaret Krieg's stress on the importance of primitive folk medicine was a vision thirty years ahead of its time.

It has only been a few years since the class of chemicals known as lectins was thought to be found only in the seeds of a few legumes. With increasing potential uses for these compounds, researchers have increased efforts to find them in other tissues and organs, not only of a wide variety of plants, but also in many lower animals.

Research and testing of new genetically engineered plants requires many years and large expenditures. What does a company do during the five to ten years the money is going out with nothing coming in? One answer to the problem has been a two step approach such as that used by DNA Plant Technologies Corporation. Bring quickly to the market new varieties based on modifications of existing breeding techniques. Assuming that the product is superior to others, cash flow will be generated to finance even better future developments.

A protein produced by the naturally occurring soil bacterium *Bacillus thuringiensis* commonly called B.t., has long been known and used for its effectiveness in controlling some caterpillars. For many years no one realized that other varieties of this organism produced insecticides harmful to other species. In conjunction with their efforts to incorporate the gene for the B.t. protein into plants to provide built-in protection, scientists have discovered a number of other B.t. compounds which are effective against some specific insects. The Mycogen Corporation is pursuing one of the most promising--a B.t. protein highly toxic to fire ants. People who do not live in areas of the South infested with the vicious uncontrollable ants cannot imagine the horror of attempting to live in their presence. In many places, children cannot play outside during the summer. The ants attack and kill pets and livestock. A frightening description of how the ants are literally taking over a Brazilian town of 7,000 residents was recently published in the **Houston Chronicle** (Blount, 1993). Mycogen's new strain of B.t. which infects fire ants might prove to be the answer to problems such as this.

The research which has produced results like these is described under the respective subjects in later chapters. The world of Awesome Green may well present us with a largesse of fringe benefits equal to those of the primary objectives.

Chapter 12

Medical Applications

This chapter is not a modern version of the Doctrine of Signatures. It is about miraculous medicines beginning to appear as a result of plant biotechnology.

The medicinal value of plants has been known since earliest recorded history. The initial uses were far from scientific. Frequently, teas made from the plant, powdered leaves, stems and flowers, or other preparations were combined with a variety of magical concoctions to relieve the patient's suffering. Later, about 4000 years ago, there is evidence that the Chinese doctors made some attempt at using measured quantities of specific preparations in the treatment of various diseases. Western medicine was slow. It wasn't until Dr. William Withering in England attempted to correlate the dosage of foxglove extract (digitalis) with the condition and specific needs of his cardiac patients. Progress continued. Various estimates point out that between one-half and three-fourths of today's drugs are directly obtained from plants or based on plant materials which have been modified for improvement.

Regrettably, history does not tell us how most of the medicinal uses of plants were discovered in early societies. This would indeed be a fascinating story. It is easy to visualize the early native American suffering from a tooth ache breaking off a twig from a willow or aspen tree and chewing on it to obtain some degree of relief from the pressure. Surprisingly, he then became aware that the pain relief went beyond that. What he was actually using, of course, was aspirin. Willows and aspens are rich in salicylic acid compounds. It is not so easy to visualize how the use of a woody gnarled geranium root as a suppository for the treatment of hemorrhoids came about. That must have been some kind of act of absolute desperation. The stories of medical applications of plants are a worthy subject for one's time. Several excellent references are listed in this book.

The new technologies to expand the medical values of plants are an equally intriguing story. It is a story which would fill a volume itself, but we shall examine here some representative aspects of this new applied science.

An early success in the modern technology occurred when a plant cell culture process was used commercially for the first time to produce an antileukemic drug. The drug was derived from a natural metabolite of the common periwinkle (*Vinca* species). The drugs known as vincaleukoblastine and vincristine were extracted from the cultured cells, purified, and used to treat countless numbers of leukemia cases. In the mid 1980's that was a giant step forward.

The technology has progressed with almost unbelievable rapidity. Such is the character of recent biologic progress. An advance a few years ago which was worthy of a Nobel prize will hardly get looked at when the experiment is repeated by a student for a high school science fair. The judge's attitude becomes, "nice project, but so what; it's old hat."

A Second Look

Science went through a period during the mid-1900's when it became popular to believe that plant sources of drugs were no longer of importance because anything needed could be synthesized artificially and more economically. By the late 1900's, we had learned that such is not always the case. As a result, today there is a mad rush by pharmaceutical companies to explore primitive medicinal applications of plant materials, particularly in tropical areas. Deforestation practices are creating many problems of survival for hundreds of species of plants that grow only in a limited range. The folklore of these regions tells of medicinal uses. If the varieties become extinct, drugs of tremendous value may be lost forever.

New technology may have to be applied to assure an adequate supply of pharmacologically active compounds obtained from endangered species, or even from uncommon species which have a very limited growth range. A good example of the latter is the development of cell culture processes to produce the potential anticancer agent known as Taxol. This compound was originally found in the Pacific Yew tree, *Taxus brevifolia*, a small tree which grows slowly under very limited geographic and climatic conditions of the northwestern Unites States and southwestern Canada. Cell culture methods were developed to obtain enough of the material for valid clinical testing. Patents for the processes involved are being issued in 1993.

Even as this book is being written, a new chapter in the story of Taxol is unfolding. Those who oppose the use of recombinant DNA may not like it, but Mother Nature has spoken in terms that cannot be misunderstood. The proudly held patents on production of Taxol by plant cell culture may turn out to be valid, but worthless.

This late-breaking discovery involves a fungus which lives in association with the Pacific Yew. Its discoverers, Drs. Andrea Stierle and Gary Strobel of Montana State University and Dr. Donald Stierle of the Montana College of Mineral Science and Technology found the fungus in the phloem of a single tree. Apparently the gene for Taxol production in the yew is also in the DNA of the fungus!

Cytoclonal Pharmaceutics, Inc. of Dallas, Texas has been licensed to commercialize Taxol production by the fungus. Dr. Arthur T. Bollon, who is chairman and CEO of this young company has had extensive research experience with fungal processes. A major goal of his current work with the organism is to obtain increased quantities of the Taxol. Dr. Bollon points out that if successful, their process will probably be more economical than plant cell production. Generally, fungus cultures can be more easily controlled than plant cell cultures.

When interviewing them to obtain more information for this chapter, the author asked both Dr. Andrea Stierle and Dr. Bollon if the presence of the plant cell gene code for Taxol in the fungus could be considered an example of natural DNA recombination. Dr. Stierle answered that at the present time she would consider that idea a working hypothesis, but definitely not yet proven. Dr. Bollon provided the same answer in only slightly different words: "It is speculative at this time that the mechanism involves genetic transfer. That is possible, but not yet proven." Much to their merit, both of these investigators showed their scientific integrity by being unwilling to make factual statements based on unproven data. Young people considering a science career will do well to emulate the approach of Drs. Bollon and Stierhle.

Both of these researchers emphasized a fact presented in Chapter 10. Recombinant DNA is nothing new and threatening. It has been occurring naturally among many organisms for countless centuries.

Once again, the opening of one door may lead to many others. Dr. Bollon feels that the discovery of the tree-fungus relationship by the Montana group may represent only the beginning of a whole new area of discoveries. He is referring to the fact that there are innumerable known relationships between various fungi and plants. The Montana investigators are now beginning a program of actively seeking fungi which live in association with plants that produce known compounds of potential medical value. Dr. Stierle points out that with modern analytical techniques it is now possible to detect natural products in very small quantities. Once they are found, and a fungus producing the same compounds can be found with the plant, the possibilities for new drugs become almost endless.

A Different Drug - A Different Method

Taxol is not by any means the only naturally occurring compound which has anti-cancer activity. Many others are known. One with much potential is the compound known as camptothecin which is an alkaloid obtained from the stem of a Chinese native tree *Camptotheca acuminata*. Its anti-leukemia and other anti-cancer activity has been known since the early 1970's. Unfortunately, in the first clinical trials in the early 1980's, camptothecin was found to have severe toxic side effects which made it impractical for use in its naturally occurring form. Artificial synthesis of the compound was achieved in 1975, but because it is a difficult and expensive process, the artificially produced material is not potentially as valuable as the natural product. As awareness of the toxicity became apparent, the problems of artificial synthesis became a moot question.

Later, scientists became aware that a relatively slight change in the chemical structure would eliminate many of the toxic side effects. Sometimes only relatively minor changes are needed. Figure 12.1 shows the chemical structure of naturally occurring captothecin. Figure 12.2 shows one of the derivatives which has the greatest promise. All it took was replacing the hydrogen attached to the number 10 carbon with a hydroxyl (-OH) group. The modified compound shown in Figure 12.2 would be known by the common name 10-hydroxy camptothecin. Chemically, camptothecin is properly known as 4-Ethyl-4-hydroxy-H-pyrano-[3',4':6,7]indolizinol[1,2-o]quinoline-3,14(4H,12H)-dione. You

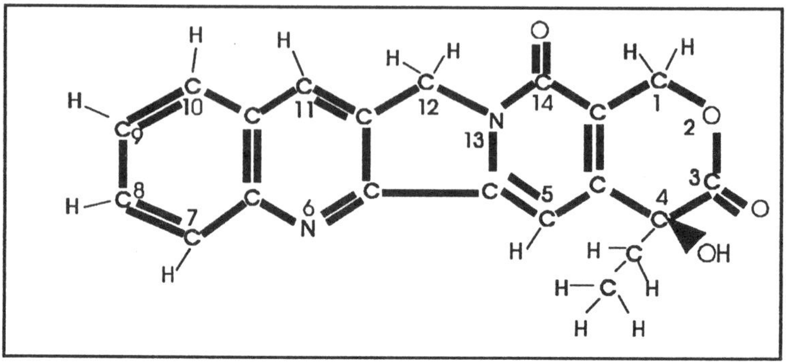

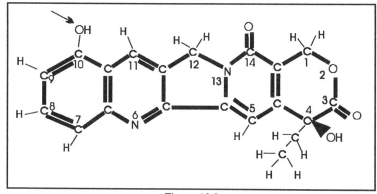

Figure 12.1

Figure 12.2

can see why it is known by its common name!

With the availability of the hydroxy form, interest in camptothecin as a therapeutic agent has been renewed. One of the problems is supply of the naturally occurring material. Dr. Craig Nessler and his colleagues at Texas A&M University have recently succeeded in producing camptothecin from cell cultures. The rate of production from the cultures makes them more economical then extraction from the whole tree. At the present time, Dr. Nessler is working to select cell culture processes which will increase the amount of camptothecin produced. He points out that the hydroxy derivative is also produced by the cell cultures, but in a quantity much too low for practical extraction. At the present time he feels that chemical modification is the way to achieve the result. The work of the A&M group at present is directed primarily to increasing the amount of camptothecin produced.

In a conversation with the author, Dr. Nessler stressed the importance of basic research to provide an understanding of the metabolic pathways and enzyme processes which result in the plant's production of a specific compound. Once these basics are understood they can be applied to a variety of products. One possibility would be to increase the amount of hydroxy camptothecin.

The group at Texas A&M is not limiting its research to camptothecin and the understanding of basic physiology processes. They are also concerned with the production of other compounds with potential anti-cancer activity from *Catherantahus rosea* and other plants.

Does some of this have a familiar ring? Once again, the important basic research must be done to provide the foundation for the application.

Science no longer exists alone in an ivory tower.

As almost every discipline has become more complex and involved, it has likewise become more dependent upon related advances. Often the glamorous and publicity demanding result of research is only the tip of the iceberg.

A good example of this is the cactus research described in Chapter 11.

The earlier knowledge that prickly pear cactus contained antidiabetic principles almost died of neglect until recently when extensive scientific investigations were started. In 1993, Dr. Eulogio Pimienta Barrios of the University of Guadalajara reported a partial characterization of the antidiabetic principle obtained from the fruits of this cactus. At the time of this writing, extensive clinical trials with prickly pear in the treatment of diabetes mellitus are under way in Mexico City. Preliminary reports (1993) by Dr. A. F. Nunar, Chief of the Department of Internal Medicine at the Instituto Mexicano Seguro Social show promising results.

Regardless of how promising and successful the work of Drs. Barrios and Nunar might turn out to be, the results will be of little practical value if a sufficient supply of the cactus is not available. If this effort does turn up a new treatment for diabetes, the real unsung heroes may well be those whose behind-the-scenes research provide the means for production of prickly pear on a large scale. Such efforts are widespread. Dr. Giuseppi Barbera and Dr. Paolo Inglese of the Universita Degli Studi in Palermo, Italy have recently (1993) provided an overview of prickly pear cultivation in their country. Similarly, Dr. Avinoam Nerd, professor at Ben Gurion University of the Negev has reported (1993) on cultural practices for cactus fruit production in Israel. In the meantime, Dr. Juan Moctezuma of Nopalili Foods, Inc. in Hidalgo, Mexico has been studying production of *Opuntia* species with irrigation under plastic. Control of cactus diseases is also a major factor if large scale production is to be achieved.

As in the case with all the new foods, development is a series of progressive steps from laboratory to greenhouse to field trials to production. Figure 12.3 shows growth of various prickly pear varieties in a greenhouse at Texas A & I University. They are being evaluated for disease resistance and herbicide tolerance.

Figure 12.3

Figure 12.4

Finally, the bottom line will be filled in by people like Robert Mick of Sinton, Texas. Mr. Mick describes himself, "I'm a farmer, not a scientist." He is, however, a practical researcher and is expending great effort in experimenting with different cultural practices for the spineless *Opuntia* species on the vast arid acres of his farm land. Figure 12.4 shows a test plot at Mick Farms. Regardless of its soundness, the theoretical results from the laboratories will serve little purpose other than solving questions of academic curiosity if practical applications cannot be made by the engineers, food processors, machinery designers, and farmers.

Eat Your Vaccination

When most people hear the word "vaccination," their immediate knee-jerk thought is likely to be "shot." Calling it a "hypodermic injection" doesn't help. It all translates into "somebody's going to stick a needle in me!" From most recipient's standpoint, the vaccine in the form of a small pill to be swallowed would be great. What about something even greater? What about a vaccine which is an inherent part of a plant food you like to eat?

In an advance far better than science fiction, botanical biotechnology is coming close to that dream.

Oral vaccines in general have not proven successful. Doctors tend to frown when they are mentioned. The reason is that the active substance called an antigen cannot get past the barriers of the stomach and the lining of the intestine. The protein structure of the antigen is destroyed or adsorbed to a surface and does not reach the part of the body where it would function. The single notable exception is the polio vaccine. Attempts to make others have resulted only in failure and frustration. New methods, however, are on the horizon.

How does a plant produce a vaccine for human use? The new technology involves incorporating into a plant a gene which will cause it to produce the antigenic protein which will bring about the desired immune response in the person who consumes it. To avoid the problem of digestive degradation of the antigen, a chemical which will protect it from attack by digestive enzymes will be included. Just eat a banana, and in addition to all its nutrients, you receive your vaccination at the same time.

Nobody enjoys getting a vaccination, but it's something we endure because the consequences of not doing so are far worse. Having the vaccine available in the form of a plant food has many ramifications. One of the reasons many diseases run rampant in developing countries is lack of vaccination. The cost of preparing, testing, and storing conventional vaccines is often prohibitive. The logistical problems of getting the vaccines into remote areas may be equally prohibitive.

Genetic engineers who specialize in vaccine development foresee many new vaccines becoming available through recombinant DNA technology during the next few years. In addition to conventional uses to prevent infectious diseases, the new vaccine arsenal will be valuable for the control of some types of cancer, heart disease, tooth decay, and perhaps even AIDS. A major concern is how these new lifesaving preventatives can be distributed and made available at affordable costs. This is precisely where vaccine production by edible plants enters the picture.

All researchers concerned with the problem agree that the plants which produce the vaccines must be ones which are readily and economically available as foods, particularly in the developing countries of the world. With different research groups attacking the problem, it naturally follows that there are different approaches.

One is that taken by Dr. Charles Arntzen who is director of plant biotechnology at Texas A&M University Institute of Biosciences in Houston. Dr. Arntzen has worked with Dr. Hugh Mason of that institution and AgriStar, Inc. and Dr. Dominick Man-Kit Lam of Lifetech Industries. Like many other researchers, their initial work was done with tobacco because tobacco cultures are relatively easy to handle in the laboratory, and probably more is known about its physiology than any other plant. Once the procedure is established for tobacco, it is relatively easy to transfer to other plants. This group recently reported highly encouraging results from experiments designed to genetically transform tobacco plants to produce hepatitis B surface antigens (Arntzen *et al.*, 1992).

Dr. Mason and his colleagues are particularly interested in incorporating the antigen into bananas. Several unique characteristic of the banana plant made it difficult to micropropagate in the earlier days of plant biotechnology. These difficulties have been overcome and the group at Texas A & M has no problems in producing an adequate supply of banana plants for their research.

The work of Arntzen, Mason and Lam is a good illustration of how opening the door to one new technology may result in the opening of many others. Dr. Mason suggests that the techniques involved in the research with hepatitis virus might be transferred with relative ease to creating vaccines against other viral diseases. One of the specific areas of interest is the Norwalk group of viruses which produce non-fatal but very annoying intestinal infections that last only a few days. Most everyone experiences discomfort from these at times. While not serious or life threatening, they are definitely debilitating and at the very least limit one's cruising range for some time.

When asked about the time required to bring a new vaccine such as these to market and make it available for general use, Dr. Mason emphasized two major areas of concern. These are the effectiveness of the vaccine and safety. Once the technology is in place for incorporating the gene for producing the vaccine into a plant, the effectiveness and safety testing will require a minimum of five, possibly ten years.

He pointed out that one reason hepatitis B was chosen for this research is the fact that the present injectable vaccine for it is the only licensed vaccine now produced by recombinant technology. The process utilizes yeast cells as the production means.

At this point many readers are probably starting to have some questions. Not the least of these might be, "How do they know when the plant is producing the vaccine antigen? Aren't these chemical molecules too small to see even with the most powerful microscope?"

One technique used for knowing is the "blot" procedure shown in Figure 12.5. This is not a poor quality picture of three tobacco leaves! It is literally a picture of three blots. The tobacco leaves used had their surfaces removed with fine sandpaper. The leaves were then placed in contact with a special nitrocellulose "paper" to produce a blot of the leaves. The blots were then stained with a solution containing antibodies to the hepatitis antigen. In areas where the antigen was present, the darker color resulted from the antigen-antibody combination. It is obvious that the leaf on the left and the one on the bottom right contained the hepatitis antigen. The barely visible blot between these two did not and, therefore, did not produce the dark color.

At this point some of you who have been around long enough to remember elementary school in the 1930's might be saying "that's high technology?" It reminds me of something we did in the third or fourth grade. We took leaves and flowers and put them on blueprint paper and held it out in the sun to make a print of the veins of the plant." Well, well. Perhaps some of this new technology is not so frightening and intimidating after all. It is just simple basic earlier knowledge clothed in complexity.

Before one can truly appreciate the difficulties of producing an effective oral vaccine, it is necessary to understand a few basics

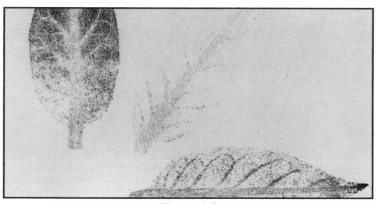

Figure 12.5

about the functioning of the immune system. Many disease
producing bacteria and viruses gain entrance into the body by
way of the digestive system. Consequently, the body's first line
of defense is found there. The lining of the intestine (as well as
other organs of the body) contains some special cells which
secrete a mucus material. Figure 12.6 illustrates the microscopic
structure of the intestinal lining. The first step in producing
immunity of this type begins when these special cells attract the
viral or bacterial protein called an antigen and transport it into
the underlying layer of epithelial cells. Among the epithelial
cells are found some immature cells of the immune system.
Some of the epithelial cells break down the bacterial or viral
proteins and the antigenic proteins released are moved to the
immature immune cells. These cells migrate into the blood and
eventually become white cells of the type known as B and helper
T cells. After these cells have matured, the majority of them
move back to the mucus surface from which they originated.
The B cells release antibody against the antigen into the mucus
at the surface of the intestine. The helper T cells join with other
white blood cells known as macrophages (from the Greek word
meaning large eaters) which attack and digest the invading
organisms. The antibodies are protein-like chemical molecules

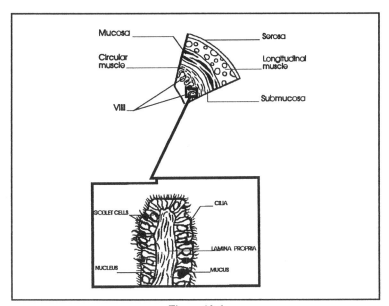

Figure 12.6

which combine with the antigen to inactivate it. Antibodies are very specific in their action. In order to work they must actually fit into a site on the surface of the antigen. Figure 12.7 shows how an antibody of one type can fit onto the surface of only the antigen which has a complementary shape to it. Not all the antibodies go to the mucosal surface. Some may remain in the blood and become part of the immune system function of *gamma globulin*.

The immunity which occurs at the intestine lining is usually not permanent. It tends to lose strength over a period of a few years unless stimulated by repeated exposure to the antigens of the disease producing organisms. This situation is ideal for vaccines in plants because the person would receive frequent re-exposure and, therefore, maintain a high level of immunity at the site where the disease producing organism would enter the body.

Every coin has two sides. The flip side of the one we have been talking about is a condition known as "oral tolerance." This term refers to the fact that by means not yet known, the body turns off the immune response to common proteins in the food we eat. That could be a major hurdle in developing a plant vaccine. Scientists involved in this research hope that the problem will be prevented by the fact that oral tolerance usually does not occur with respect to proteins which are part of a larger structure such as a virus. Therefore, causing the plant to produce the antigen as part of another chemical structure might be the answer.

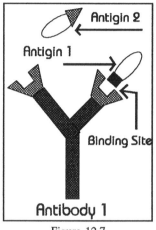

Figure 12.7

Once again the house cannot be built before the foundation is in place. This is simply another example of how applied research must be supported by knowledge obtained from basic research.

Building on the initial basic research with tobacco, Arntzen and his group have extended their work to the successful transformation of lettuce and tomatoes to produce the vaccine component. These researchers are now evaluating a number of plants for use in vaccine production. Their emphasis is on food varieties which are commonly produced and used in tropical regions where most of the developing countries of the world are located. Other scientists are concerned with other diseases.

Another likely candidate for plant vaccines is cholera, a severe and often fatal intestinal infection spread by contaminated food and water. It is a major disease in many tropical and semitropical countries. Public health officials in highly developed nations live in constant fear of an outbreak of cholera following natural disasters such as floods and hurricanes that would leave water supplies untreated and spread sewage contamination widely. Dr. Michael Hein and the group he heads at the Scripps Institute of Research in La Jolla, California are finding promise of success with alfalfa.

In the last two years, Hein's group has succeeded in transforming alfalfa plants to produce a nontoxic portion of the cholera toxin as an antigen in alfalfa sprouts. Mice fed with these sprouts have shown immunity when attempts to infect them with cholera were made (Hein *et al.*, 1992). The researchers are now in the process of attempting to grow enough of the transgenic alfalfa to provide material for more extensive testing. In the meantime, they are investigating other plants as possible subjects for the vaccine production.

One plant which is drawing attention from all researchers involved with edible vaccines is the soybean. While no success has yet been reported with it, extensive research is involved because soybeans are a major part of the food supply in many developing countries.

Nutraceuticals

Don't feel dumb if you did not recognize that heading for this section. It is another new word - one recently coined by Dr. S. L. DeFelice. He used it to describe foods that contain substances which provide disease prevention or therapeutic health benefits. Many of these foods are not new; they have been used for thousands of years in Chinese and other traditional medicines. Now, they are drawing scientific attention and their medically active components are being identified.

The benefits of many traditional medicines are long established, but generally have been ignored by western medicine. One of the problems is that the salutary effects may lie in natural combinations of several chemicals in the food. Western pharmacology wants a single chemical.

In many instances the single specific is well-known. The effects of Vitamins C and E as antioxidants are widely recognized as is the efficacy of *beta* - carotene in retarding some types of cancer.

Less well-known in the United States is the effect of extracts of *Ginkgo biloba* leaves in counteracting cerebral insufficiency in the elderly. Various teas and other preparations of this nutriceutical are extensively used in Europe to combat diseases, memory failure, and confusion.

The Chinese are now attempting to verify the activity of many traditional herbal remedies. For example, an English translation, A Barefoot Doctor's Manual, of a Hunan Province medical manual for lay practitioners, describes chemical and pharmacologic evaluation of some plant extracts.

As specific active ingredients become known, there should be no problems in transforming more common foods to fabricate the desirable chemical. This would be a logical approach for a company such as DNAP which promotes its new produce varieties as "Value Added" products.

Chapter 13

Disease Resistant Plants

While not clothed in the glamorous attention getting trappings of anticancer drug research, the development of new disease resistant plant varieties is of vital importance. In some cases, such as Dr. Wambugu's research with sweet potatoes, it means extending the growth range of a needed food. In other cases, the value is more one of economics--less costly production, better yields, and ultimately lower costs to consumers. A major result will also be reduced use of pesticides. Plants that are naturally disease resistant do not need treatment with pesticides, such as fungicides or nematocides. All classes of economic poisons are capable of creating major health problems not only for the consumers, but also for the producers and handlers throughout the distribution system. A shocking example of the latter is the thousands of cases of pesticide poisoning among farm workers in Thailand reported in the early 1990's. Another, involving farm workers in Latin America, is discussed in Chapter 16.

Many of these "new" genetically engineered plants that will become common in the 21st century are not really all that new. They are simply modern enhancements of folklore and practices that have been common among farmers and home gardeners for generations.

Even before the advent of chemical pesticides, plant growers were aware of the use of many natural biological controls. One approach is what was called "companion planting." Many old-timers realized that if they had a plant such as a rosebush that was infested with mildew or other fungus problems, planting garlic or onion nearby would often eliminate the disease.

Although they probably did not realize it, they were simply taking advantage of some sulfur containing compounds emitted by these plants. The sulfur compounds were toxic to the mold which caused the disease of the rose.

Other approaches involve the use of friendly insects such as the Praying Mantis which will consume many harmful insects and not itself bother the plants. As knowledge and technology developed, other approaches were used.

A good example of this is the marigold. Members of the marigold family have long been known to produce a nematode repellant chemical in their roots. In fact, many years ago a variety which produced more than the usual amount was marketed under the trade name Nemagold™. Most of the nematocides and nematode repellant compounds produced by the family Asteraceae are derivatives of complex chemicals called polyacetylenes. If the genes for these compounds could be combined into the genome of other plant species, they conceivably would be able to produce identical compounds in their own root systems. This would be a major advantage for many food and forage crop species as well as ornamentals. Treatment of growing plants for nematode infestation is difficult in their environment. There are problems of penetration of the nematocidal compound into the soil and problems of toxicity of the best nematocides to other forms of life. The naturally occurring production of repellents by the root system would obviate such concerns.

Choose Up Sides

Once the appropriate genes have been identified, it is not difficult to incorporate them into a plant to cause the plant to produce defensive compounds against disease organisms. Everyone agrees on that. The scientific debate starts with the question of how the plant should behave with respect to the new compound. Should it be produced constantly, thereby giving the plant continuous protection, or should it be introduced into a metabolic pathway which will function like those of many naturally occurring phytoalexins? In the latter case, the plant does not actually produce the compound until it is attacked and some initial damage has occurred. In other words, the attack is the trigger that sets off the defense. Many scientists believe this is the best objective. They argue that the plant will not be wasting its energy and resources constantly producing a compound that might not be needed. The opposing argument is that if the compound is constantly present, the plant is less likely to suffer any damage at all from invading viruses, bacteria, or fungi. This could well turn out to be one of those debates in which both sides are partially correct and both are partially wrong. The future may reveal that one method is better for certain plants, while the other would be more desirable for different ones. Time will tell.

B.t.

Bacillus thuringiensis, commonly called B.t., is a widely dispersed natural soil bacterium which produces a crystalline protein that is toxic to many caterpillars. B.t. has been used for many years in controlling various caterpillar predators of different plants. It is undoubtedly the safest insecticide known because the organisms does not infect most other insects or higher animals including man. When the host worms die, the bacteria die. Mass cultures of the bacteria have been used as a dust or spray in controlling many plant diseases since the early 1960's. Unfortunately, growing enough of the bacteria for widespread use is an expensive proposition. Scientists began to think in terms of finding a better way.

After the structure of the bacterial product toxic to the worms was discovered, its genetic code was considered for recombination into some plants. Since the material which is toxic to the worms is harmless to humans and all higher animals, incorporating it would provide the plant with built-in protection. Several companies are now involved individually and in joint efforts in attempts to incorporate the gene for the *B. thuriengiensis* toxins into a number of plants. Two serious problems which may be the first to be controlled by this process will be the corn borers and the tomato hornworm.

Different strains of B.t. produce their effects on different insects. They are very specific with regard to the insect host and are totally harmless to other insects and higher animals.

The pictures on the next page show the striking results of the incorporation of the gene for producing B.t. proteins in two major crops. Figure 13.1 and 13.2 were taken on the same day in adjacent fields with cotton plants of the same age. The first picture on the left is of normal cotton which has been infested by the cotton bowl worm. The plants below were protected by the B.t. gene which had been incorporated. The potato plants in the center row of Figure 13.3 contain a B.t. gene for producing a protein harmful to the Colorado Potato Beetle. The decimated plants on either side do not contain the gene. Equally striking is the corn in Figure 13.4. Both plants were cut lengthwise down the middle. The plant on the left shows corn genetically improved to resist the European corn borer. The results of attack by this insect can be seen in the stalk at right from a plant that did not contain the B.t. gene.

Can farmers plant seed for these improved plants now? No, not at this time. The plants shown are from extensive field tests being conducted by Monsanto Company scientists to establish safety and effectiveness of the transgenic varieties. Monsanto expects them to be available for general use between the mid and

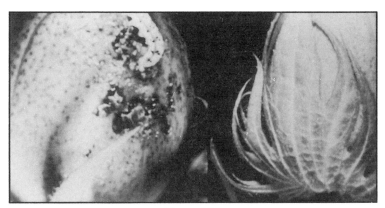

Figure 13.1

Figure 13.2

late 1990's. One might ask with results like those shown why wait? The Monsanto scientists, like other plant biotechnologists are a conservative group. Contrary to the picture painted by some anti-technology activists, the industry is vitally concerned with safety as well as function. They want to be sure. They are also carefully fulfilling the safety and efficacy requirements of various federal and state regulatory agencies.

Figure 13.3

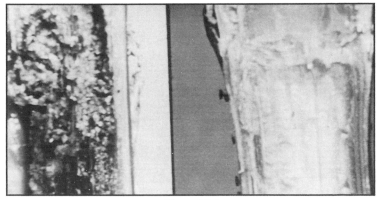

Figure 13.4

(Photos 13.1-13.4: Courtesy the Monsanto Company)

The Monsanto Company is joined by others in evaluating the genes from different strains of B.t. for protection of a wide variety of different plant species and cultivars. The research of all is at about the same stage of development.

An important spin-off of such genetic engineering research is the ability to improve delivery of what we already have. B.t.

proteins have been used for some time with a perfect safety record as a spray or dust on many crops. A problem with such applications is the rapid chemical or biodegradation of the compound when it is applied. The Mycogen Corporation has come up with an improved product for current types of application. They call it Cell Cap™. Figure 13.5 is a photomicrograph of B.t. crystals as they appear when isolated from the bacteria. Mycogen likes to refer to these as the "naked" crystals, the ones which degrade rapidly when applied. Figure 13.6 shows similar crystals enclosed in a biodegradable protective Cell Cap™. The covering does not prevent the crystals from being effective when ingested by a target insect. It does help extend the time of effectiveness of the application. This technology could well eliminate the need for many potentially harmful pesticide applications. Even the most vociferous anti-technologists who are constantly searching out possible problems have never objected to B.t. insecticides.

Mycogen's research efforts have resulted in the isolation of many previously unknown strains of B.t. Some of these may prove to be highly effective against pests which are now difficult to control. One of the most interesting of these really has nothing to do directly with plants, but since the research that led to its discovery was the result of plant technology, it is certainly worthy of mention. A recently isolated strain produces a compound which is toxic to fire ants (*Solenopsis saevissima*). If more research continues to support the initial findings of safety and effectiveness, people who live in areas infested by the ravaging ants might well consider this discovery to be the most important of all related in any way to plant biotechnology.

Herbicide Resistance

Overgrowth of weeds is as much of a threat to desirable plants as microbial diseases. This, hopefully, provides the justification

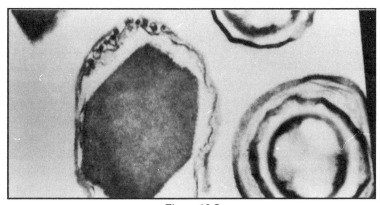

Figure 13.5

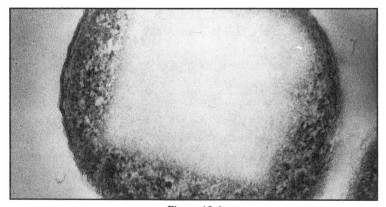

Figure 13.6

(Photos 13.5-13.6: Courtesy Mycogen Corporation)

for including this subject in the chapter on disease resistant plants.

The value of genetically engineering plants to be more herbicide resistant so that herbicides can be used more freely is the single area of the new plant biotechnologies that the author seriously questions. In all honesty, the question reflects a strong personal prejudice which has occurred as a result of others improperly or even illegally using herbicides in a way that damaged the author's property as described in Chapter 16. At the same time, it must be admitted that if the herbicides had been used properly according to their labels, the damage would not have occurred. We shall try to be objective.

Many of the earlier herbicides such as 2,4,5-T and its by-product, the infamous Dioxin of Agent Orange of Vietnam, did cause major environmental problems. Human birth defects resulting from spraying in the Pacific Northwest are well documented. Some herbicides have long-term soil residues or may be easily washed or carried great distances from their point of application. Anything that can be done to eliminate such compounds is a step in the right direction. A major step is being made by genetically engineering plants to be resistant to the herbicide glyphosate properly known by its chemical name of N-(phosphonomethylglycine)isopropylamine salt. This compound is produced and sold under the trade name "Roundup" by the Monsanto Company. It is a broad-spectrum contact herbicide which is non-selective and will kill plants to which it is applied through their foliage. The compound is absorbed by the leaves and moves throughout the rest of the plant, gradually resulting in death. From the environmental standpoint, one of the best features of Roundup™ is that it has no activity in the soil. Therefore, it cannot wash or otherwise be distributed to affect nearby vegetation to which it has not been directly applied. Being able to use such a compound will perhaps eliminate the need for many others that are potentially more harmful, not only to other plants but also to humans and higher animal life.

Plants which produce an enzyme known as EPSP Synthase have the ability to break down glyphosate before it can harm them. The single gene for the production of the enzyme is now known and has been incorporated into a number of major crop plants on

Figure 13.7

(Courtesy The Monsanto Company)

an experimental basis. Figure 13.7 is a picture of Canola plants containing the gene along with normal plants between them. All were sprayed with Roundup. Other crops being investigated for herbicide resistance include soybeans, cotton, oil seed rape, and corn. Monsanto scientists hope that these crops will have passed all required safety tests and become available by the end of the 1990's.

The genetic engineering of economically valuable plants to resist a desirable herbicide is a worthwhile achievement. It will be reflected in lower tillage rates and lower use rates and fewer applications of other chemical controls which might be environmentally detrimental.

"Stay Away Or Get Dissolved"

If plants could talk, some might soon be warning insect and fungus predators with the words above.

The external skeleton or "shell" of most insects and the cell wall of many fungi are composed of a complex, dense carbohydrate material known as chitin. Some decomposer bacteria produce an enzyme known as chitinase which breaks down an insect's shell. Plant biotecnology is studying ways of transposing plants by giving them the gene for chitinase production. If it works, the insect or fungus may start to dissolve as soon as it takes the first bite of the plant!

One of the worst and most widespread soil fungi pathogenic to plants is *Rhizoctonia solani*. It causes a variety of root rots and stem infections. When soil, moisture, and temperature are ideal for it, *R. solani* can destroy the young plants of a major crop. In some parts of the U.S., it causes major problems in lawn grasses. The diseases are difficult or expensive to control with ordinary fungicides.

The growing tips of the fungus are highly sensitive to chitinase. Although many plants have a gene for making chitinase, they do not produce it in defensive quantities until triggered by a fungus attack. For young, tender seedlings, that is often too late.

Scientists at E.I. DuPont Company are developing a means of giving plants the ability to produce and store chitinase at all times so the plant will have protection even at the beginning of the fungus invasions (Broglie *et al.*, 1991).

They identified and modified a gene which causes some varieties of beans to produce chitinase without an infection onset. Using *A. tumefaciens* as described in Chapter 10, they experimented with tobacco which is highly susceptible to *R. solani* caused diseases. The tobacco plants incorporated the chitinase gene. Eighteen day old seedlings with and without the gene were transplanted into *R. solani* infested soils and grown for about two weeks. The mortality rate of the transgenic plants was only approximately half that of the normal controls which had a death rate of 53 per cent. Transgenic plants producing greater amounts of chitinase had even lower death rates.

As usual, experiment with tobacco and then move the application to economically or nutritionally valuable crops. Similar results were obtained when the technique was applied to canola plants. Other likely major crop candidates will be corn and soybeans.

Chitinase provides a highly specific type of protection, affecting only fungi which have chitin in their cell walls. Other pathogens such as *Pythium aphanidermatum*, which does not have chitin, were unaffected in the DuPont tests.

The DuPont team points out that even if the effects of the chitinase gene are expressed only in young seedlings, a major goal will have been accomplished because, in general, more mature plants are less affected by *R. solani* diseases.

The research is another striking example of how transforming a plant with a single disease resistance gene may eliminate the need for chemical pesticide application.

Human safety questions about these transgenic plants is utter nonsense. First of all, human and other higher animal cells do not have chitin; secondly we have been consuming chitinase in our plant foods since time began.

Chapter 14

Industrial Products

The products of the new plant biotechnologies are by no means limited to food and drugs. Many materials with promising industrial use lie just beyond the horizon.

Most of us have experienced what we thought were plastics in plants - those vegetables that have spent too long on the cafeteria steam table. Now, the real thing begins to appear!

One of the most novel ideas from the new technology is obtaining plastics from plants. This has already been accomplished on a limited scale by Dr. Chris Somerville and his associates at Michigan State University (Poirier *et al.*, 1992). The plastic they produced is known as polyhydroxybutyrate (PHB). This plastic resembles polypropylene in many respects. It is strong, somewhat brittle, and highly resistant to attack by many chemicals. One big advantage it has over polypropylene is that the PHB is easily biodegradable. Only a brief contact with the soil is required for enzymes produced by common soil bacteria to break down the plastic. It is not now obtained as are many other plastics from petroleum products. The relatively small amount in use today is produced by a bacterial fermentation process. Dr. Somerville and his group have succeeded in transferring the genes responsible for the plastic production from bacteria into plants. Their initial success was achieved with a variety of cress.

As with many other new processes, the initial success was accompanied by some problems. The plants apparently devoted much of their energy to producing PHB instead of normal quantities of their compounds required for growth; consequently, the plants expressing the PHB genes were stunted and abnormally small. Some ranged from only 19 to 45 per cent of the size of the natural controls. However, Dr. Somerville is not particularly concerned about the problem because, as he points out, the ultimate goal would be to achieve PHB production in much larger plants which are adaptable to full scale agriculture production. Two likely candidates are non-edible varieties of sugarbeets and potatoes. If the same growth problems occur in the larger plants, Dr. Somerville feels that they can be overcome by manipulating the plant's metabolism so that it will grow normally while producing large quantities of PHB.

Another problem was the fact that PHB formation is not a simple single step process. Actually, three major steps are involved. Consequently, three different genes control formation of the enzymes needed. Only one of these occurs naturally in the experimental plants. Therefore, two different bacterial genes had to be included in the final transgenic plant in order for it to carry out the complete pathway:

Natural Precusor Compound $\xrightarrow{\text{Enzyme 1 (Natural)}}$ Intermediate Compound 1 $\xrightarrow{\text{Enzyme 2 (Recombinant)}}$

Intermediate Compound 2 $\xrightarrow{\text{Enzyme 3 (Recombinant)}}$ PHB

You can see that if this pathway is interrupted at any point, PHB will not result.

Seeds obtained from some of the transgenic plants were fertile and carried the new genome which was expressed in their offspring.

According to Dr. Somerville, PHB is not the only plastic which could conceivably be produced in plants.   Others are being experimentally produced by various bacteria, and the gene coding for them could be transferred to plants.

Oils

The production of oils by plants is nothing new. For years man has used natural oils extracted from plants for a variety of food and industrial purposes. The most common include corn oil, soy oil, and peanut oil. Coconut oil finds a use in the making of high quality soaps, other cosmetics, and industrial products. The same thing holds true on a more limited scale for palm oil and olive oil.   Castor oil produced by the castor bean plant is reported to have over four hundred different industrial uses. This may come as a surprise to those whose only contact with it has been when it was used as a purgative.   A major problem associated with castor oil is that the plant which produces it is essentially tropical in nature.   Its lack of low temperature tolerance limits the range in which it can be grown. One goal of the research with castor oil is to transfer its genetic code from the castor plant into others such as the sunflower which grows under temperate conditions.

Another oil of interest, both for food and industrial purposes, is Canola oil.   It is produced in the seeds of the canola plant. Compared to more common oils and fats used in food processing and cooking, canola is somewhat expensive. The higher cost is out of proportion to the benefits that can be gained from its unique characteristics.   Monsanto and several other biotech companies are actively involved in research programs to transfer the gene for the oil from the canola to other plants which might produce it on a larger scale, thus making it more economical. Other efforts are directed toward improving the oil produced by reducing the percentage of saturated fatty acids it contains.

While a major use of canola oil at present is in frying potato chips, many other uses exist. An increased supply would result in new applications for cosmetic, lubricant, and various other industrial formulations.

Lignin

Lignin is a non-living strengthening material produced by some plants to give added support to their cells. It is part of what makes some plants stiff and "tough." The genetic control of lignin production is becoming known. Someday genetic engineering will be able to reduce the amount of lignin in some species, making them softer and more biodegradable. Other plants might benefit from added lignin production to make wood materials stronger.

Chapter 15

The Foundations

Tomatoes that taste like tomatoes
Plants producing plastics
Hepatitis vaccine in lettuce
Food for developing nations
Treatment of leukemia

That is the glamour and the glory--the end result we are aware of and see. It is the completed 2000 piece jigsaw puzzle. It was by no means a pretty picture when the box was first opened. What about all those pieces - how were they made? How much tedious work was expended in selecting the right ones and putting them together? All too often the basic pieces are overlooked as we observe the final result.

Commenting fairly about the basic research which went before is an almost impossible task because of the sheer numbers of contributors. Those such as Haberlandt and White who laid the early cornerstones were followed by almost countless others who found the little bits and pieces that made the end result possible. From the 1970's through the later 1980's, hundreds of researchers supplied the knowledge of plant life processes which were necessary for the present manipulations. Developing a new stain to make plant organelles more visible under the microscope, identifying a previously unknown enzyme, discovering that protoplasts fuse more readily in the presence of ethylene glycol--these are just a few examples of the underlying knowledge which had to be in place before applications could be made.

In many cases, the research which provides the knowledge base for later application may not even seem related. A good example of this is the literally hundreds of research reports concerned with various aspects of movement of materials through the membrane which surrounds a cell's contents. This research was done over many years on both plant and animal cells. Knowing the structure of the membrane and how it functions to permit or block passage of materials through it was essential before the techniques of genetic engineering could be developed and applied.

Another question often asked is "Why has so much of the research effort in plant biotechnology been devoted to tobacco?" The answer to that question is quite simple. Tobacco (*Nicotiana* species) is relatively easy to work with under laboratory conditions. It grows rapidly, and early workers used it to develop the base of knowledge from cell and tissue culture. Other researchers built upon this base until a body of knowledge was available. Once the functions of its system were understood and the techniques were in place, transfer to other plants was easy.

Think about a central theme of biology to which these examples point. It is the basic similarity of life forms and processes. Only a very few DNA sequences make the difference between a rose and a turnip. The rest are alike. In fact, a large number of DNA codes and their resulting functions are identical in both animals and plants. While all species of life have their unique distinguishing characteristics, the basic processes are identical. The process of respiration in plants follows the same cycle as it does in man.

Sometimes the line between basic and applied research is not clearly drawn. A good example of this is the work of Dr. Park Nobel at UCLA that was cited in an earlier chapter. His discovery that water content of cactus is related to the temperature at which freeze damage occurs is certainly a practical application for *Opuntia* species producers. At the same time the knowledge of the relations of water functions to freezing in cactus might be applied to many problems which seem unrelated at this time. Who can predict how or where these basic principles of function might be used five or ten years from now?

The questions of freezing in living tissues cross the boundary lines between the plant and animal kingdoms. Marine biotechnology is interested in the mechanisms which protect some fish species from freezing at temperatures which would kill others. One of the early discoveries is an "antifreeze" protein which interferes with ice crystal formation. Transforming some fruits and vegetables to produce this protein might prevent them from becoming mushy when frozen and thawed. Another possibility is to incorporate the antifreeze into cold-sensitive plants to extend their growing range.

Test Tube Fertilization

Even as the final manuscript for this book was being prepared, a significant advancement in fundamental research was reported. Earhard Kranz and Horst Lörz (1993) at the University of Hamburg achieved for the first time a test tube fertilization with isolated single plant sperm and egg cells. Zygotes produced by their technique developed zygotic embryogenesis and finally resulted in fertile heterozygous maize plants.

The initial reaction to this might well be "so what - test tube fertilization has been done for years." Yes, with animal gametes, not plants. The process of fertilization of a plant ovum is much more complex than that of animal cells. In animals, the sperm and egg function totally independent of other cells and tissues. While the plant sperm, once it has formed from the pollen grain, functions with relative independence, the egg cell does not.

Intricate relationships between the egg cells and some somatic cells which are associated with them have precluded *in vitro* fertilization.

The technique described by Kranz and Lörz will undoubtedly open many doors to new knowledge and new products over the next few years. Our basic knowledge of early plant embryology has been very limited by the fact that associated tissues have before been necessary for embryo development from the zygote. The fact that this is no longer needed will provide a basis for filling the void in our knowledge of these essential processes.

This new basic knowledge will undoubtedly be a springboard for many significant applications. One of the first applications predicted by Kranz and Lörz will probably be transformation of the *in vitro* zygote by microinjection or electroporation with foreign DNA. This has the potential for resulting in an almost unlimited wide variety of new or modified plants.

When reading the report of this research, little imagination is required to visualize the countless hours of tedious work required to achieve their notable success. The senior researchers freely acknowledged that this success was the result of teamwork by many individuals, including laboratory technicians, assistants, and greenhouse growers who brought the resulting plants to maturity. As the world becomes more specialized in its niches of knowledge, such group effort will characterize most significant research achievements.

Another important basic advance involves DNA. A single DNA molecule has the code for many different genes. Only those which apply to a particular cell are expressed in that cell. The others are turned off. For example, the genes which control the shape of one's finger are also present in the nose. Your nose (hopefully) does not look like your finger because the finger genes are not expressed in the nose. They are only expressed in the fingers.

Controlling the expressions of specific genes has been a major problem in the early development of plant biotechnology. The fact that a desired "foreign" gene can be spliced into the DNA of a plant cell does not mean that the product coded by that gene will be produced.

The discovery that gene expression is controlled by another nearby DNA region called a "promotor" provided the answer to the problem of causing the cell to produce the desired material. Genetic engineering is now beginning to combine promotor sequence with the desired gene to get it expressed. Such combinations are called **chimaeric genes** (Burnett and Nessler, 1993). Figure 15.1 illustrates how such DNA might look.

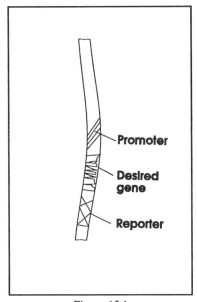

Figure 15.1

Another special type of gene called a reporter gene is controlled by the promotor gene. The picture below shows how the promotor activity can be visualized by a reporter system which produces compounds that produce a blue color when an enzyme controlled by the reporter gene splits a special dye.

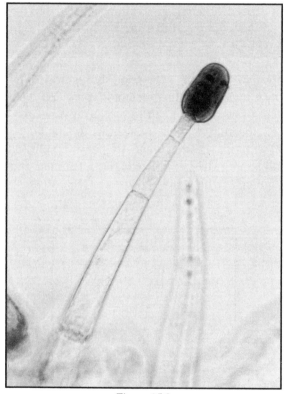

Figure 15.2

The dark pyramid-shaped structure is a tobacco leaf trichome showing evidence of a promotor by the dense color. Trichomes

are outgrowths of epithelial cells. They have a variety of forms ranging from hairs to glandular structures like the one shown.

Being able to direct and control specific gene expression and visualize the results of it provides a major basic tool for many different applications in plant biotechnology research.

Society will find itself in serious error if it allows the products of applied research to overshadow the basics which made them possible. Failure to support basic research may not have an immediate effect, but a few years down the road nations which do not support it will find themselves sadly lagging behind in the world wide competitive market place.

Take Notice

Chapter 16

Safety--Does Anybody Care?

Yes.

The anti-technology activists have performed a superb job of planting seeds of fear and concerns of the safety of genetically engineered new foods. The biotechnology industry has not effectively countered the suspicions created in the public mind. Safety is still a major worry.

Most consumers are little aware that many of our common foods are derived from species which have toxic properties. Among others, this group includes tomatoes, potatoes, celery, and squash. Some of these, tomatoes most extensively, were used in ancient times specifically for their toxic effects. Because of their hallucinogenic and toxic traits, tomatoes were touted by the ancients as an aphrodisiac and were widely used to promote licentious sexual behavior. Whether they actually did or not is open to question, but the toxic effects are well known.

Science Questions Safety

Because the plants listed above have long been known to contain genes for possibly toxic substances, breeders have always been very cautious about checking new varieties produced by conventional techniques. This concern is continued in evaluating new varieties produced by biotechnology processes. There is basically no difference in what we are doing now compared with what has been done for many years.

The May, 1992, decision by the Food and Drug Administration that bioengineered foods do not need to be labeled and require no special testing beyond that required for any new variety produced by older cross breeding techniques set off a storm of protest. Critics raised ridiculous questions of fear and safety. The FDA rightfully pointed out that new varieties produced by modern technology basically are no different from the standpoint of safety than those produced by conventional breeding techniques for hundreds of years.

Contrary to the ideas promulgated by some of the special interest groups, the FDA is not ignoring safety by not requiring every new food variety to be submitted for approval and labeling. Restrictions do apply. One of great importance is that any new variety which contains a potential allergen must have prior investigation. Similar restrictions apply to all foods such as those named above which have varieties with known toxic properties. Additionally, if a new food variety contains a protein which is unusual or unique in some way, that protein might be considered a food additive under the regulations although it is being produced naturally by the plant. In such cases labeling would be required.

Ignorance Questions Safety

"Ignorance" here is not used in a derogatory sense; rather, it implies not knowing. Much of the concern about bioengineered food plants arises as a result of lack of knowledge and understanding about the processes involved. Our society does not need more stress. What it does need is to accept the fact that we live in a technological age and that this new technology must be understood from the early periods of a child's education. It is no longer justifiable nor socially and educationally acceptable for elementary teachers to say, "I don't understand science." They must understand science, and they must communicate it to their students. In today's world, we would question whether they have any business teaching if they do not understand science. In a recent interview by **Scientific American Magazine** (September 1992) Alvin Young, Director of the USDA office of Agricultural Biotechnology stated, "From a scientist's point of view a gene is just a functional unit that causes a certain reaction." His profound understanding of the question might need to be emblazoned on a large sign over the entrance door to every college of education.

Those who have managed to stay with us through 143 pages of this review so far are well aware by now that the author has a generally favorable opinion of the scientists involved in biotechnology. They deserve an A for their technological achievements. Now, lets go down the report card to the category labeled "Communication." Here we will have to hand out the F. Well, maybe it can be a D-. You know, the kind of D- a teacher gives when the student really deserve an F, but because there were one or two bright spots, decided to give the D instead for encouragement.

The scientists understand the problems and potential dangers that might exist. In general, they have taken knowledgeable conscientious precautions to avoid problems and dangers, but unfortunately their concerns often do not reach the consumer. A good example of this is the review of metabolic engineering of plant products by Dr. Craig Nessler of Texas A & M University. At one point in this lengthy highly technical report, Dr. Nessler states, "More importantly, however, this data underlines the fact that one cannot assume that only the expected products of specific enzyme reactions will accumulate in transgenic plants transformed by a specific enzyme. Rather, a detailed chemical analysis of a number of individual transformants must be made to empirically determine the outcome of each metabolic engineering experiment."

Dr. Nessler, a professor of biology, has raised the caution flag and pointed out possible pitfalls in proper scientific terminology in this review which will soon be published in **Transgenic Research**. You will not find this publication at your local news stand, but it is in the research libraries of major universities and corporations. Dr. Nessler is addressing other scientists, not the general public. He has done his job and carried out his responsibility. Communicating such scientific care to the potential consumers should be the responsibility of the corporations which will ultimately profit from it.

Unfortunately, they seem to have failed to understand the lack of technical knowledge of the general public and, as a result of this lack of awareness, have failed to communicate the soundness of plant genetic research. Like it or not, the public's knowledge of biotechnology in general and genetic research in particular is derived from science fiction books and movies.

Obviously, many of the companies involved in plant genetic research are not in a position to throw money around carelessly. They are in many cases funded by venture capitalists and other risk takers. Nevertheless, they might do well to spend a few thousand dollars on public relations and public education efforts to avoid spending millions later as a result of lack of public understanding. Those involved in the prickly pear research described earlier provide an example that others might study and emulate. Dr. Felker has become actively involved with scientific laymen in presenting his cactus research. He communicates freely with farmers, ranchers, food technology and marketing people, and others concerned. They, in turn, have presented it to the public in a positive manner. As a result, his research has evoked no fear; to the contrary, it has received highly favorable treatment by the major popular press such as the **Houston Chronicle** (July 1991). Effort and willingness to communicate even with those who do not speak your native tongue--perhaps this is what distinguishes the truly successful from all the others.

Economic Poisons

Those who raise questions about the safety of genetically engineered plants might do a greater service by directing their efforts toward what we are eating now. Genetic engineering deals usually with a single gene related to the plant's metabolism and having no relation to human physiology. One gene is controllable. What about the thousands that may be transferred unknowingly in conventional cross breeding techniques? What about the pesticide residues on the foods we now consume? The **Houston Chronicle** recently featured a frightening report (Lambrecht, 1993) about pesticide residues on fruits and vegetables imported from Latin America. The same author also called attention to pesticide poisoning of farm workers in developing countries - an almost unbelievable number estimated by the World Health Organization at 25 million cases. Safety is present in agriculture? Is not the risk lessened if we have a new variety made naturally resistant to insect predators so that it requires no treatment with insecticides? Even if the insecticides that remain on edibles are within well established tolerance limits, the question of concern is what is the long term effect of accumulation of a multitude of these "tolerable" levels?

Another factor for consideration is that pesticides do not always remain were they are placed. A recent example is the concern expressed by public health officials about pesticides washed from farmlands during the midwest floods of 1993. Some of these were carried hundreds of miles from their point of application and the use for which they are registered.

The pictures on the next page are graphic examples of how an application can go wrong as a result of carelessness on the part of the user. Under terms of the right-of-way agreement with the pipeline company, the owner of the land shown has full agricultural use of the surface. Without his knowledge or

Figure 16.2

permission, the company erected the sign in the middle of a rare and endangered classic rose bush. Herbicide was applied to the rose. Within a few hours, a heavy rain occurred and washed the herbicide many feet along a drive as shown in Figure 16.2. Safety in present practices?

Many dangers involved in current practices are accepted without questions. One of the worst is aerial applications of insecticides and herbicides. Again it is not the gun that kills--it is the person that pulls the trigger. Aerial applicators are afforded special flying privileges by the FAA. Unfortunately, some do not exhibit the special sense of responsibility which should accompany that privilege. The application often occurs within the legal windspeed limits established by a weather observer several miles from the application point; at that point there may be wind gusts which will spread the poison beyond the limits of the field to which it is being applied. Some economic poisons can become problems even when legally and properly applied. Ask anyone who lives within a few miles of a rice field. What happens the next day after a herbicide is legally applied? During the heat of the following day part of the substance vaporizes, and the vapors may be carried many miles by the wind. With the cooling of evening they condense and fall upon property miles away. Safety in present agriculture?

Another danger is residues of the pesticide which adhere to parts of the aircraft and later fall or are blown off miles from the point of application.

Perhaps it is time to put the protests where they need to be--on the reality, not on imagined unlikely fears associated with a new technology and new words. Our environment stands only to gain if this new technology results in new varieties which require the less of economic poisons.

"Nature still offers her bounty,
and human efforts
have multiplied it."

FRANKLIN D. ROOSEVELT
Inaugural Address

Part IV

Report Card Time

Chapter 17

"But, Mom, Look At All the A's"

One of the great pleasures of doing research for a book of this type is the opportunity for the author to make many new acquaintances in academia, business, and industry. Another was that researching this subject in depth revealed that the potential of this new world of awesome green is even greater than originally thought.

Unfortunately, a few clouds occasionally cast their shadows. This author has never been known for reluctance to express his opinion, even when that opinion might result in taking shots at some of the holy cows of the education establishment (Tant, 1991) such as the theorists who meddle in science teaching although they know nothing about science. There is no apparent good reason to change now, so here goes.

If you look at a single painting by an artist, you certainly are aware of its characteristics. Your awareness does not go beyond that single work. However, if you examine many paintings by the same artist, a motif of style, talent, and interest begin to emerge. Such has been the case resulting from conversations with many in the biotech industry. The pattern that struck the author is that many of this young industry's problems seem to be ones related to shortcomings in public relations.

Somewhat surprisingly, the problem appeared in industry and not in the colleges and universities where the author's reception, as noted in the foreword, was universally cordial.

Even in business and industry there were some bright spots. The outstanding PR program of those involved in the cactus research has already been described. Another bright spot was provided by a young small company, Cytoclonal Pharmaceutics Inc., which seems to have a clear understanding of the need for public knowledge. Mycogen Corp. is another company that is conscious of the need for communication.

Many of the companies contacted gave the impression that they were suffering from a severe case of paranoia--somebody was out to get them if they so much as showed their faces or opened their mouths. It is not hard to understand how such an attitude might have developed when one considers the number of times they have been fired upon and kicked by the anti-technology activists. Although a caution light may shine, it is time for the industry to grow up and start behaving like the reputable and trustworthy companies they actually are. In many instances, there may also be a factor of erroneous or incomplete descriptions in the popular press written by writers who have no science background. Sometimes all the author could get was a canned spiel mouthed by a PR representative whose understanding of the subject obviously did not go beyond a prepared text. Certainly, a research scientist has things to do other than spending all his time talking to writers. Nevertheless, there are occasions when a short technical conversation would be of mutual benefit. If a company lacks confidence that its research scientists will not give out proprietary information, it might be time to look at replacing those scientists. If a question concerning such matters is asked, a simple statement, "I'm sorry but that is proprietary information" or "I'm sorry I can't comment on that right now because we are in the process of preparing a patent application on it." would suffice. The writer would understand and the company would still receive favorable treatment in the press.

Who would think that the public relations department of some companies would not return telephone calls? On two occasions this unbelievably occurred even when the PR person's name had been ascertained beforehand and the call was placed personally to those individuals. Both these cases involved companies that are large enough and have been around long enough to know better. Perhaps it is time that someone does create a mutant monster plant and feed it with people who do not return phone calls!

Some companies expressed pride in their efforts to assist in biotechnology education. Unfortunately, they may not be aware that their efforts have in many cases been to cooperate with some national or regional groups which get the information to curriculum supervisors, but not to the front lines. The teachers who would use the information are never aware of it.

The author became curious about teachers' name recognition relative to the plant biotech industry. An informal, totally unscientific, and totally statistically invalid poll was done by asking 20 high school science teacher acquaintances to name five companies in plant biotechnology. None were able to provide either four or five names. Two gave three names. Only four were able to provide three names. Eight came up with a single name. Seven were unable to name any.

Regardless of its lack of validity, the brief survey sounds a warning. These were not just average teachers. They are all in a large metropolitan area. They participate in science fair sponsorship and other activities which help teachers stay current. Some are science department chairs. Here is a list of companies and the number of times they were named:

Monsanto: 7
Dow: 5
E. I. DuPont: 4
AgriStar: 2
Calgene: 1
General Foods: 1
Nabisco: 1

Could there have been some guessing or other factors involved? Draw your own conclusion: the two who named AgriStar live near it. Several of those who named Dow and Monsanto live near major chemical production facilities of those companies.

Free unrequested advice is usually worth exactly what one pays for it, but here is some for the biotech firms anyway. If you don't want the science fiction mentality view of biotechnology to continue into the next generation, today is the time to get in direct contact with science teachers and get some into your labs to see what is going on. Notice, we said *teachers*, not curriculum directors or other administrators who do not know a centrifuge from an electrophoresis cell. Someone might even have to work on a Saturday so teachers can come. Don't worry about that--teachers are accustomed to working on Saturdays grading papers or preparing labs for the next week. Find ways to help teachers keep up to date. Teachers who are not located near large universities rarely have access to current scientific journals. The textbook publishing and distribution process is such that by the time a new textbook is ready for the schools, the information in it may be two years old. In a rapidly developing technology that makes it useful for a history lesson.

Another major problem for the biotechnology industry is money. Help the public become aware of the costs involved in commercializing a new product. Many people think of plant biotechnology as being the domain of such corporate giants as Monsanto or DuPont. They have little awareness of the hundreds of new small companies which are struggling to make the new technology available. When that new product does finally get to market, you will be much less likely to receive accusations of price gouging if the consumer has some idea of the research and development costs involved. Closely related to this, in fact an integral part of it, is the time required to bring an experimental product to the consumer. The average person is little aware of the years required for evaluation and safety testing. In most cases he is not at all aware of the myriad government regulations which must be met.

The governmental regulatory requirements are enough to make biotech company management seek a comparative haven of order in an asylum for the insane. Some new products will have to meet requirements of the FDA, EPA, and Department of Agriculture. In many cases, they must also comply with different state laws. It is not unusual for regulations at different agencies to be in conflict with each other. Ask your congressman why, but don't expect an answer.

More knowledge in these areas would also help offset the public conception fostered by some popular writers that the biotech industry is consumed by greed and avarice and has no consideration for safety factors.

Let the public know that the genetic engineers are not a group of weirdos from outer space. Likewise, they are not wild-eyed Dr. Frankensteins embarked on a course of creating monsters that will ultimately reap destruction. The average citizen needs it know that scientists are people not unlike all others. They have children, help sponsor little league, attend PTO meetings, and participate in church or community functions like anyone else. Neglect of basic public relations principles has often helped the anti-tech activists paint a false image of the people involved in this new field of research.

Another problem may have its roots in our education system. Professions and training have become highly specialized, but plant biotechnology encompasses many areas. We need to make certain that biologists, chemists, engineers, and producers can communicate with each other to achieve common goals.

In any rapidly expanding field of the scope of the new plant technologies, there are always going to be some rough spots. The problems here are manageable, and except for conflicting government regulations, they should be easily overcome. There is too much good and too much potential for wonderful discoveries benefiting all mankind for the industry to stumble over these few areas.

Welcome to a new world of truly awesome green.

Epilogue

In her fascinating book **Green Medicine**,*
Margaret Kreig quoted from the field notes of a plant explorer:

The Different Plant

"I wonder what's around the bend?"
said the explorer.
"I wonder what that plant is?"
said the collector.
" I wonder what's in it?"
said the chemist.
" I wonder what activity it has?"
said the pharmacologist.
"I wonder if it will work in this case?"
said the physician.
" I hope she lives!"
said the father.
" Please, God!"
said the mother.
" I think she'll be all right in the morning."
said the nurse.

Life. That is really what this book has been about.

* Copyright 1964, Rand McNally Co.

Selected References

Anonymous. 1992. A storm is breaking down of the farm. *Business Week* **December 14, 1992**:98-99.

Bailey, James E. 1991. Toward a science of metabolic engineering. *Science* **252**:1668-1674.

Barbara, Giuseppe and Pado Inglese. 1993. An overview of prickly pear production in Italy. *Proc. Fourth Ann. Prickly Pear Conv.*:(in press).

Blount, Jeb. 1993. Brazilian Village Trying to Stamp out fire ants. *Houston Chronicle* **Sept. 19**:A24.

Broglie, Karen, I. Chet, M. Holiday, R. Cressman, P. Biddle, S. Knowlton, C. J. Mauvais, and R. Broglie. 1991. Transgenic plants with enhanced resistance to the fungal pathogen Rhizoctonia solani. *Science* **254**:1194-1197.

Burnett, Ronald J. and Craig L. Nessler. 1993. Crop improvement through directed cell-specific gene expression. *AgBiotech News and Inf.* **5(2)**:75N-78N.

Chuck, George, T. Robbins, C. Nijjar, N. Courtney-Gutherson, and H. Dooner. 1993. Tagging and cloning of a petunia flower color gene with the maize transposable element activator. *The Plant Cell* **5**:371-378.

Evans, David A. and William R. Sharp. 1988. Tissue culture of Lycopersicon spp. *U.S. Patent 4,734,369.*

Evans, David A. and William R. Sharp. 1986. Applications of somaclonal variation. *Bio/Technology* **4**:528-532.

Evans, David A. 1988. Applications of somaclonal variation. *In Biotechnology in Agriculture*:203-223. Alan R. Liss, Inc., NY.

Feitelson, Jerald S., Jewel Payne, and Leo Kim. 1992. Bacillus thuringiensis: insects and beyond. *Bio/Technology* **10**:271-275.

Fraley, Robert. 1992. Sustaining the food supply. *Biotechnology* **10**:40-43.

Frati-M., Alberti. 1992. Medical implications of prickly pear cactus. *Proc. Third Ann. Prickly Pear Conv.*:29-30.

Frati-M., Alberti. 1993. Clinical trials with prickly pear cactus and diabetes. *Proc. Fourth Ann. Prickly Pear Conv.*:(in press).

Freeman, Karen. 1993. Monsanto and Kenyan scientist target improved sweet potato. *Gen. Eng. News 13* 7:1.

Gruwell, Charles E. and Frank H. E. Preene. 1937. Antidiabetic substance. *U.S. Patent 2,082,952.*

Guha, S. and S. C. Maheshwari. 1964. Production of plants from anther culture. *Nature* **204**:497.

Jorgensen, Richard A. and Carolyn A. Napoli. 1991. Genetic Engineering of Novel Plant Phenotypes. *U.S. Patent 5,034,323.*

__, __. 1993. Genetic Engineering of Novel Plant Phenotypes. *U.S. Patent 5,231,020*, continuation in part of U.S. 5,034,323.

Kotl, K. and W. Kasha. 1985. In Bright, et al, Cereal Tissue and Cell culture. *Martinus Nijhoff-W. Junk, Publishers.* Dordrecht, The Netherlands.

Koziel, Michael G., G. Beland, C. Bowman, N. Carozzi, R. Crenshaw, L. Crossland, J. Dawson, N. Desai, M. Hill, S. Kadwell, K. Laumis, K. Lewis, D. Maddox, K. McPherson, M. Meghji, E. Merlin, R. Rhodes, G. Warren, M. Wright, and S. Evola. 1993. Field performance of elite transgenic maize plants expressing an insecticide protein derived from Bacillus thuringiensis. *Bio/Technology* **11**:194-200.

Kranz, Erhard and Horst, Lörz. 1993. *In vitro* fertilization with isolated single gametes results in zygotic embryogenesis and fertile maize plants. *The Plant Cell* **5**:739-746.

Krieg, Margaret. 1964. Green Medicine. *Rand, McNalley Co.*, Chicago.

Kunitake, Hisato, Hideo Imamizo, and Masahiro Mii. 1993. Somatic embryogenesis and plant regeneration from immature seed-derived calli of rugosa rose. *Plant Sci.* **90**:187-194.

Lambrecht, Bill. 1993. Tainted produce pours into U.S. despite inspections. *The Houston Chronicle* **Nov. 14, 1993**:A24.

Liebert, Mary Ann. 1993. Get an Early Start. *Gen. Eng. News 13* **12**:4.

Lu, Chin-Yi. 1993. The use of thidiazuron in tissue culture. *In Vitro Cell Dev. Biol.* **29P**:92-96.

Mason, Hugh S., Dominic Man-Kit Lam, and Charles J. Arntzen. 1992. Expression of hepatitis B surface antigen in transgenic plants. *Proc. Nat. Acad. Sci. USA* **89**:11745-11749.

Morrison, Robert A. and David A. Evans. 1988. Haploid plants from tissue culture: New plant varieties in a shortened time frame. *Bio/Technology* **6**:684-690.

Morrison, Robert A. and David A. Evans. 1991. Pepper Gametoclonal Variation. *U.S. Patent 5,066,830.*

Morrison, Robert A., David A. Evans, and Zhegong Fan. 1991. Haploid plants from tissue culture: Application in crop improvement. *In Subcellular Biochemistry* **Vol. 17**:Plant Genetic Engineering, B. B. Biswas and J. R. Harris, Eds. Plenum Press, NY.

Nadakavukaren, M. V. and Derek McCracken. 1985. Botany - An Introduction to Plant Biology. *West Publishing Co.*, Minneapolis, MN.

Nerd, Avinoam and Y. Mizrahi. 1992. Effects of fertilization on prickly pear production in Israel. *Proc. Third Ann. Prickly Pear Conv.*:1-8.

Nessler, Craig. 1984. Metabolic engineering of plant secondary products. *Transgenic Research*:(in press).

Nobel, Park S. 1994. Remarkable Agaves and Cacti. *Oxford University Press, NY*:(in press).

Nobel, Park S. and M. E. Loik. 1993. Low temperature tolerance of prickly pear cacti. *Proc. Fourth Ann Prickly Pear Conv.*:(in press).

Oeller, Paul W., Lu Min-Wong, L. P. Taylor, D. A. Pike, and A. Theologis. 1991. Reversible inhibition of tomato fruit senescence by antisense RNA. *Science* **254**:437-439.

Perez, Facundo, Fernando Borrego-Escalente, and Peter Felker. 1992. Collaborative Mexico/United states initiative to breed freeze tolerant fruit and forage Opuntia varieties. *Proc. Third Ann. Prickly Pear Conv.*:49-55.

Poirier, Yves, Douglas E. Dennis, Karen Klomperans, and Chris Somerville. 1992. Polyhydroxybutyrate, a biodegradable thermoplastic, produced in transgenic plants. *Science* **256**:520-523.

Rust, Carol. 1991. Cactus crop. *Texas Magazine of the Houston Chronicle* **June 6**:6-8.

Schneider, Elizabeth. 1986. Uncommon Fruits and Vegetables. Harper and Row, New York.

Stephanopoulos, Gregory and Joseph J. Vallino. 1991. Network rigidity and metabolic engineering in metabolite overproduction.*Science* **252**:1675-1681.

Taylor, Robert. 1993. Research Briefs, Food for Thought:"Seropositive" plants may yield cheap oral vaccines. *Jour. NIH Res.* **5**:49-53.

Thorpe, Trevor A., Ed. 1983. Plant Tissue Culture Methods and Applications in Agriculture. *Academic Press*, NY.

Weber, Patricia. 1992. Designer biopolymers from recombinant DNA technology. *Science* **258**:39.

Weymann, Kris, K. Urban, D. M. Ellis, R. Novitzky, E. Dunder, S. Jayne, and G. Pace. 1993. Isolation of transgenic progeny of maize by embryo rescue under selective conditions. *In Vitro Cell Dev. Biol.* **29P**:33-37.

Young, Alvin. 1992. In Science and Business, Deborah Erickson, Ed. *Scientific Amer.* **Sept.**:160-162.

The New Language--A Glossary Of Terms

Plant biology has changed. New theories and concepts, new facts, and new processes have resulted in a proliferation of new technology. To anyone who has not yet had the opportunity to follow recent developments in this technology, the new language is a confusing foreign tongue. New abbreviations are being rapidly coined for the new words and processes. Unfortunately, some editors of scientific publications are permitting the use of these abbreviations without proper definition.

All this makes it even more difficult for busy persons to keep abreast of the new developments. If we are unable to communicate our new knowledge, can it have any real and lasting value?

Glossary

Abscisic acid: A plant hormone responsible for leaf and fruit drop; involved in the control of normalcy and also inhibits seed germination.

Adventitious: Refers to the development of plant organs from unusual origin such as callus tissues.

Alleles: Members of a pair of genes.

Androgenesis: The development of a haploid male plant from a pollen grain.

Aneuploidy: The condition in which the number of chromosomes differs from the haploid number or multiples of the haploid number.

Antibodies: Substances usually proteins produced by an organism as a defense against foreign materials.

Antigen: Foreign substances which illicit an antibody or immune response.

Antisense DNA: A backward copy of the DNA code for a peptide.

Apomixis: Refers to different types of asexual reproduction substituting for sexual reproduction. It implies no gamete fusion occurring.

Asexual: Reproductive processes which do not involve two parents.

Axenic: Refers to a pure culture consisting of only a single type of cell or organism uncontaminated by others with differing properties.

Browning reaction: The occurrence of a brown color in freshly cut tissues or tissue culture media usually as a result of the production of phenolic compounds.

Callus: An undifferentiated group of cells that forms as a response to plant tissue injury or hormone imbalance.

Cell cycle phase (G phase): The series of events in a cell which results in the formation of two daughter cells. "G" is sometimes used to indicate specifically a growth phase following cytokinesis.

Cell suspension culture: A tissue culture technique involving the multiplication of cells suspended in a liquid medium.

Chimera: A plant which has two or more different genomes. As a result of the dissimilar genes, the plant may have phenotypically differing distinct appearances.

Chloroplast: Chlorophyll containing plastids.

Chromoplast: Plastid containing photosynthetic pigments of colors other than green.

Chromosome: The structure which contains the DNA of a cell.

Clone: A genetically identical group of cells, tissues, or plants. Common usage of the term implies that all members originated from a single common source.

Complete flower: A flower which contains the nonessential floral parts of sepals and petals in addition to the reproductive structures.

Crossing-over: The exchange of corresponding cell segments between members of a pair of chromosomes.

Cryopreservation: Refers to various methods of preserving tissues by use of liquid nitrogen or other low temperature processes.

Cuticle: A waxy protective coating produced on the leaves of many plants.

Cybrid: A cytoplasmic hybrid which originates from the fusion of a cytoplast with a protoplast or intact cell.

Cytokinens: A group of plant hormones which stimulate cell division; also involved in formation of side shoots and breaking dormancy in some seeds.

Cytokinesis: The division of the protoplasm which ordinarily follows mitosis.

Cytoplast: The part of a cell consisting of the organelles surrounded by the plasma membrane but not including the nucleus. The word is sometimes misused as a synonym for protoplast.

Dedifferentiation: The loss of specialized functions and features as a cell reverts to a meristematic or callus state. It is sometimes caused by injury or hormone imbalance.

Dicot: The major class of flowering plants characterized by embryo with two seed leaves, net leaf veins, and flower parts in multiples of four or five.

Differentiated: The development of specialized cells and tissues having specific functions from undifferentiated meristematic, callus, or proembryonic cells.

Dihaploid: Refers to a cell or organism which arises from a tetraploid. It is distinguished by a chromosome number of 2n=2x in contrast to the tetraploid condition of 2n=4x.

Diploid: The condition of chromosomes in which they exist in pairs; characterize of all non-reproductive cells.

Electroporation: Enlargement of pores of a cell membrane caused by a sudden surge of electric charge.

Embryogenesis: The formation and development of an embryo; sometimes misused to refer to the development of a plantlet from an embryo.

Embryoid: An embryonic plantlet produced artificially from somatic cells *in-vitro*.

Endomitosis: A condition of polyploid nuclei following division of the chromosomes without separation of the chromatids to form daughter nuclei.

Endopolyploidy: The production of polyploid cells following endomitosis in some tissues of a higher plant.

Endosymbiosis: A symbiotic relationship in which one of the organisms lives within the other to their mutual benefit.

Etiolated: A condition of abnormal chlorophyll development and excessive elongation of the internodes usually resulting from plant growth with inadequate light.

Euploidy: The normal stable condition of the number of chromosomes being double the haploid number.

Explant: The part of a plant removed from the parent and used to initiate *in-vitro* culture.

Gametophyte: The sexually reproducing generation of a plant

Genome: All the DNA which is the code for every inherited characteristic.

Genome: The total number of genes ie., the genetic make-up of the organism. It is usually considered to be the haploid number of chromosomes.

Germplasm: All the genetic reproductive material of an organism.

Habituation: The condition in which after a number of subcultures, plant tissues adapt to grow without hormones or other specialized nutrients originally needed.

Haploid: Used to describe the condition of a cell which contains only one member of each pair of chromosomes.

Hardening off: The acclimatization of plants produced *in-vitro* to adapt them to *in-vivo* conditions by providing an opportunity for the development of leaf cuticle or other responses.

Heterokaryon: A cell which has two or more different nuclei as a result of cell fusion.

Homokaryon: A cell which has two or more identical nuclei as a result of cell fusion.

Homoplastic hybrid: A cell resulting from the fusion of two identical cells or protoplasts.

Illuminance: The visible radiation striking a surface.

Imperfect: A flower which has the reproductive structure of only one sex.

In-vitro: A term used to describe biological processes occurring under artificial laboratory conditions.

In-vivo: A term describing biological processes occurring in living organisms.

Incomplete flower: A flower which does not have all the essential and nonessential structures.

Intergeneric: Used to describe a cross between two different genera.

Interspecific: Used to describe a cross between two different species.

Irradiance: The total radiation, including visible and invisible, falling on a surface.

Karyogamy: The fusion of two nuclei during sexual reproduction, or the fusion between two nuclei following plasmogamy.

Lectin: A glycoprotein of non-immunoglobulin nature which causes erythrocyte agglutination.

Leucoplast: A plastid which does not function in photosynthesis.

Macro: Large; usually implies that it can be seen without magnification.

Macrophages: Large defensive cells which consume foreign invaders such as bacteria and viruses.

Meiosis: The process by which pairs of chromosomes divide so that only one member of each pair will appear in each egg or sperm produced.

Mericlone: Usually applied specifically to orchid clones resulting from meristematic multiplication *in-vitro*.

Meristem: The rapidly dividing cells in the growth tips or other regions of a plant.

Metabolic engineering: Artificially changing the normal chemical processes of a plant in order to achieve the production of unusual products.

Micro: Small; usually implies needing magnification to be visible.

Microballistic injection: A process for inserting DNA into a cell by firing microparticles of metals coated with DNA through the cell membrane.

Microinjection: The use of an microscopic needle to inject materials into a single cell.

Micropropagation: Cloning or other vegetative propagation of plants *in-vitro*.

Mitochondria: The respiratory apparatus of a cell, involved in oxygen and carbon dioxide exchange.

Mitosis: **Replication of the DNA and division of a cell nucleus as the initial step before cell division.**

Monocot: One of the two major classes of flowering plants; characterized by a single seed leaf, flower parts in multiples of three, and parallel leaf veins.

Mutagen: Any substance or condition which will increase the normal mutation rate.

Nucellar embryo: An embryo which develops vegetatively from the somatic tissues surrounding the sexually produced embryo.

Organ culture: The production of an organ *in-vitro* using techniques which cause it to be similar to the original organ from which it derived.

Organogenesis: The development of specialized organs from undifferentiated cells.

Ovagenesis: The development of a haploid organism from an unfertilized egg or other female reproductive structures.

Ovary: The female reproductive structure of a plant in which the eggs develop.

Ovule: The female reproductive structure which matures into an egg of an plant.

Perfect: A flower which contains the reproductive structure of both sexes.

Petal: An individual unit of the chorea of a flower usually large and often brightly colored in insects polluted species.

Photoperiod: The light phase of alternating light and dark periods.

Photosynthesis: The process by which a plant converts light energy into chemical energy.

Phytoalexin: The name given to a class of chemical compounds produced by plants as a defense against disease. The word "alexin" means to ward off. Originally thought to prevent fungal invasion.

Phytohormone: Regulatory chemicals produced in plants.

Pistil: The female reproductive structure of a plant.

Plasmids: Small circular structures of DNA found in some bacteria.

Plasmogamy: The fusion of haploid protoplasts during sexual reproduction.

Plastids: The photosynthetic structure of a plant.

Plating efficiency: Refers to the percentage of inoculated cells which develop into cell colonies.

Polyembryony: The condition in which two or more embryos form after fertilization.

Polyhaploid: A plant which has half the number of chromosomes of a polyploid of that species.

Polymers: Large chemical molecules composed of hundreds or thousands of repeating units.

Polyploid: A plant which has multiples of its normal base number of chromosomes as triploid or tetraploid.

Primary culture: The initial growth resulting from an explant of cells, tissues, or organs.

Primordium: A cell or group of cells which develop into a tissue or organ.

Propagule: A plant or plant part used to initiate cloning or other propagation.

Proplastid: A plastid-like rudimentary structure thought to mature into a plastid.

Protoplast: All of the contents of a plant cell inside the cell wall.

Recombinant DNA: Deoxyribonucleic acid containing unusual gene sequences which have been obtained and inserted from a different variety of cell.

Regeneration: The regrowth or replacement of differentiated tissues or organs which have been removed from a plant.

Rejuvenation: A return to an immature juvenile state.

Reverse transcriptase: An enzyme which transfers the code for a nucleotide sequence from mRNA to DNA.

Runner: A specialized reproductive stem.

Secondary product formation: The production of phytochemicals from suspension cell cultures or other *in-vitro* techniques.

Sepal: An individual unit of a flower claylx, which is the leaf like structures covering and protecting developing buds.

Single node culture: A culture developed from tissue taken from one node of a stem.

Somaclonal variation: The increased genetic variability which often occurs with repeated *in-vitro* culture.

Somatic hybridization: The development of a hybrid plant from non-reproductive cells.

Somatic embryogenesis: The development of embryos in somatic cells such as leaf mesophyll cells.

Spores: Specialized asexual reproductive cells formed by some plants and lower organisms.

Stamen: The male reproductive structure of a plant.

Stigma: The receptive outer end of the female reproductive structures; designed to catch pollen.

Style: The tube extending from the plant ovary to the sigma.

Tuber: A specialized food storage stem usually containing reproductive buds such as the "eyes" of an Irish potato.

Totipotent: The potential of a cell to develop into any cell type or regenerate a complete plant as a result of having the total genome of the plant present.

Transformation *in-vitro*: The occurrence of hereditary changes during the culture of protoplasts, cells, tissues, or organs.

Undifferentiated: Cells such as meristem which have not developed into specialized types of cells or tissues.

Vegetative reproduction: Multiplication of a plant by means not involving sexual fusion, eg. rooting stem cuttings.

Vitrification: A term applied to any plant showing symptoms of physiological disease.

Abbreviations

AA- Amino acid.

ABA- Mucopolysaccarides responsible for the ABO blood group system; also, Abcissic acid.

BA- Benzyl adenine.

CMS- Cytoplasmic male sterility.

pCPA- para-chlorophenoxyacetic acid.

2,4-D- 2,4-dichlorophenoxyacetic acid, a phytohormone used as a weed killer.

DNA- Deoxyribonucleic acid.

cDNA- Complementary DNA produced from an RNA template by the action of RNA-dependent DNA polymerase (reverse transcriptase).

pDNA- Plasmid DNA.

rDNA- In general, any DNA regions that code for ribosomal RNA components.

tDNA- A group of seven genes of the Ti-plasmid that integrate into the nuclear DNA of the host plant during tumor induction.

ELISA- Enzyme Linked Immunosorbent Assay

GA- Gibberellic acid.

GMO- Gamborg, Miller, & Ojima media.

IAA- Indole acetic acid.

2iP- 2-(isopentenyl)-adenine.

K or **KIN**- Kinetin.

MS- Murashige & Skoog.

NAA- Naphthalene acetic acid.

PAGE- Polyacrylamide gel electrophoresis.

PBS- Phosphate buffered saline.

PCV- Packed cell volume.

P protein- Phloem specific proteins formed following injury.

RIA- Radioimmunoassay.

SH- Schenk & Hildebrandt medium.

TC- Tissue culture.

Ti- Tumor inducing. Usually refers to a plasmid from *Agrobacterium tumefaciens*.

Science Supply Sources

Benz Scientific
P.O. Box 7022
Ann Arbor 48107
(313) 994-3880

Nasco
901 Janesville Avenue
Ft. Atkinson, WI 53538
1-800-558-9595

Carolina Biological Supply Company
2700 York Road
Burlington, NC 27215
1-800-334-5551

Nebraska Scientific
3823 Leavenworth Street
Omaha, Nebraska 68105
1-800-228-7117

Central Scientific Company
3300 Cenco Parkway
Franklin Park, IL 60131-1364
(708) 451-0150

Science Kit & Boreal Laboratories
777 East Park Drive
Tonawanda, NY 14150-6784
1-800-828-7777

Connecticut Valley Biological Supply
P.O. Box 326
Southampton, Mass. 01073
(413) 527-4030

Sargent-Welch Scientific
7400 North Linder
Skokie, IL 60076
1-800-727-4368

Edmund Scientific, Co.
101 East Gloucester Pike
Barrington, NJ 08007-1380
(609) 573-6250

Sargent-Welch Scientific, Canada
77 Enterprise Drive North
London, Ontario N6N 1A5
1-800-325-3010

Flinn Scientific
P.O. Box 219
Batavia, IL 60510-9906
(708) 879-6900

Synthephytes
P.O. Box 1032
Angleton, TX 77516-1032
(713) 369-2044

Further Reading

Angier, Bruce. 1986. Field guide to medicinal wild plants. *Stackpole Books*, Harrisburg, PA.

Pierik, R. L. M. 1987. *In-vitro* culture of higher plants. *Martinus Nijhoff Publishers*, Boston.

Reader's Digest Association. 1986. Magic and medicine of plants. *Reader's Digest Assn., Inc.*, Pleasantville, NY.

Riha, J. and R. Subik. 1981. The illustrated encyclopedia of cacti and other succulents. *Octupus Books, Ltd.*, London.

Plant Sources

Kraatz Cactus Nursery
U.S. HWY 77
Riviera, TX 78363
(512) 296-3678

Mick Cactus Farms
P.O. Box 403
Sinton, TX 78387
(512) 364-1431

Related Books By Carl Tant From Biotech Publishing

Plant Biotech Lab Manual

No more boring botany labs. Twelve experiments that can be done with a minimum of equipment in advanced high school or introductory college labs. Experience the excitement of embryogenesis, protoplast fusion, cloning, chemical extraction, and more.

ISBN 1-880319-03-9 $19.95 Comb Bound

Teacher Pack including blackline reproduction masters for reports and other class materials. $8.95

Seeds, Etc...

Seed experiments, demonstrations, and projects for grades 5-9.

ISBN 1-880319-01-2 $13.95 Softcover

Science Fair Spelled W-I-N

Exciting, unusual projects for grades 7-12. Includes some based on plant biotechnology.

ISBN 1-880319-02-0 $14.95 Softcover

Projects: Making Hands-On Science Easy

Helps for teachers and parents who have no experience with science project work. Especially applicable for grades 1-8.

ISBN 1-880319-06-3 $12.95 Softcover

Available from major book sellers or direct from the publisher. If ordering direct, please include $2.00 shipping and handling.

Index